Variable Stars

VARIABLE STARS

by John S. Glasby, B.Sc., F.R.A.S.

Constable London

First published in 1968
by Constable & Co Ltd
10 Orange Street London WC2

Printed in Great Britain
by The Anchor Press Ltd, Tiptree, Essex
SBN 09 456200 8

To my Wife

Contents

Foreword

For many decades the discovery and study of variable stars was the province of the amateur astronomer and even today, when more and more of this work is being carried out by the professionals with the largest telescopes and most up-to-date equipment in existence, the dedicated amateur is still providing valuable contributions to our understanding of the nature of these stars. The task of following the light variations of the long period variables, the semi-regulars, the 'eruptive' variables and the novae remains firmly in the hands of the 'back-yard' astronomer using only modest equipment.

This book is designed to give an overall picture of our present knowledge of the many different types of variable stars. It is not meant to be the last word on this extremely complex subject. Astronomy never professes to give the last word. It grows by a tentative exploration of the great realm of the unknown by expounding theories which, however attractive and well put together, must inevitably be modified as our knowledge grows and eventually satisfy observation, for this is the one criterion by which every astronomical theory stands or falls.

So it is with the subject of variable stars. Only in the case of the eclipsing variables do we know the true cause of the light variations, many of which are far more complicated than originally thought. For the many thousands of other variable stars we can only put forward reasoned guesses which may, in some instances, still be very far from the truth.

Holwarda can scarcely have foreseen, in 1638, the tremendous chain of discovery and research which has resulted from his finding that Mira varies in brightness in a period of about 330 days. Today, much of our knowledge of the size of the galaxy and the distances to the extragalactic nebulae was accumulated by the study of the Cepheids, the RR Lyrae stars and the long period variables. The eclipsing binaries have provided us with much vital information concerning the masses, diameters and densities of a wide variety of suns, information which it would have been very difficult to obtain in any other way and those most spectacular of stars, the novae and supernovae, are yielding important knowledge of the nuclear processes taking place in the centre of stars.

This book is based in part on a series of lectures I have given over the past four years but by far the greater proportion of it derives from the published work of specialists in the various fields of variable star observation and that of amateur observers all over the world.

In particular, it is a great pleasure to thank Professor F. Hoyle, F.R.S., for permission to quote from his work on the Algol variables and supernovae and to reproduce Figure 19, the University of Chicago Press for permission to use several of the diagrams from the *Astrophysical Journal* and the British Astronomical Association and the American Association of Variable Star Observers for use of many thousands of observations made by their members over the past fifty years.

Stevenston, Scotland J.S.G.
February 1968

Introduction

At first glance it might appear that all stars visible in the heavens remain at the same brightness from one night to the next. Indeed, for almost 2,000 years since Hipparchus and Ptolemy compiled their star catalogues giving the magnitudes of the naked eye stars, this belief does not appear to have been challenged. Since the latter half of the sixteenth century, however, it has been recognised that this is not the case, that there exist a large number of stars which vary tremendously in their brilliance, some in an extremely regular manner, others less regularly, while at the other end of the scale there are stars which are completely unpredictable and irregular in their light variation.

Thanks to the ingenious observational techniques which have been developed since the invention of the telescope and spectroscope, we are now able to investigate the behaviour of these variable stars thereby obtaining a vast amount of information concerning their physical and chemical constitution and the mechanisms responsible for the observed variations in their light.

Early in October 1572, a spectacular new star appeared in the constellation of Cassiopeia and was carefully studied by Tycho Brahe. When at maximum, this remarkable star outshone Venus and was easily visible in broad daylight but within a few weeks it had begun to fade and a year after its discovery it had fallen to sixth magnitude and was only just visible to the naked eye. Unfortunately this was some years before Tycho Brahe's observatory was built and his description of the star's position is not sufficiently accurate to enable present-day astronomers to locate the star with certainty since there are several stars, all extremely faint, close to the designated spot.

Some twenty-four years later, in 1596, Fabricius discovered the variability of a star in Cetus, the constellation of the Whale. That this was not a nova was proved by Holwarda in 1638 who, by careful observation over several years, showed that the star varied periodically in brightness in about 330 days and gave to it the name of Mira, the Wonderful, a name which it retains to this day.

Mira was later recognised as the prototype of a very large class of variable star known as the long period variables and characterised by a wide variation in brightness from maximum to minimum and periods in excess of ninety days.

More than a century later, the periodic decreases in the light of the naked eye star Algol, β Persei, which had been noted by Montanari in 1669, were correctly interpreted by Goodricke on the basis of an eclipsing binary system. The two stars making up the system are not intrinsically variable but their orbital plane lies so nearly parallel to our line of sight as seen from Earth that with each revolution about their common centre of gravity, they mutually eclipse each other resulting in a diminution in the combined light of the system. To date, these eclipsing variables are the only class for which the cause of variability is known with certainty. It is quite possible that even before the discovery of the variable nature of β Persei by Montanari, early Arab astronomers were aware of its periodic dimming since the name Algol may be a corruption of the Arabic – al-ghul meaning The Demon.

Naturally, so long as astronomers were hampered by a lack of suitable optical instruments, the number of variable stars discovered remained small. Towards the middle of the nineteenth century, however, a systematic search for these objects began with Argelander and was ably continued by many professional and amateur astronomers. At first, the method used was to compare star fields with the charts prepared several years before, in particular the Bonner Durchmusterung which gives the positions and magnitudes of all stars down to visual magnitude 9.5 and any stars which were not recorded therein were considered to be possible variables and observed closely for any change in brightness. Towards the end of the nineteenth century photography was employed, two plates of the same part of the sky being examined under the blink microscope when any stars which had changed in magnitude during the intervening period were readily recognised. This technique had the tremendous advantage that a large number of stars could be examined at the same time.

So rapidly has this technique been developed that at the present time more than 25,000 variable stars are known and several hundred are being added each year. Many of these are regularly observed in order that the type of variability may be determined and to provide data from which we may someday discover the cause of the light and spectral variations.

It was early realised that variable stars form a very special class and accordingly, although they are designated by a letter or letters, followed by the name of the constellation as with other stars, the capital Roman letters have been reserved for them beginning with R for the first variable to be discovered in any constellation, followed by S to Z. Doubled letters are then used from RR to RZ, then SS to SZ, TT to TZ, etc. Once ZZ has been reached the nomenclature continues with AA to AZ, BB to BZ and so on until the series is exhausted with QZ. This series allows for a total of 334 variables in any particular constellation and the nomenclature is then V335, V336 and so on. The capital letters A to Q could not be used for variable stars since these had already been utilised in the naming of non-variable stars in the newly discovered southern constellations. It should be noted however that certain of the brighter variables which had already been allotted Greek letters by Bayer have had these retained, e.g. β Persei, o Ceti (Mira) and δ Cephei.

From careful and continuous observation of the light changes of a variable star it is possible to plot the magnitude against time and thereby construct a light curve by means of which we are able to classify these stars into several fairly well-defined types. During their light cycles, changes also occur in the spectra of these variables which are characteristic, to a certain extent, of the class of variable with which we are dealing. For example, the Cepheids, named after δ Cephei, the brightest star of this very important class have spectra of types B, A, F and G whereas the long period variables are almost exclusively of types M, N and S. From this we can tell that the former are hot, white or yellow stars while the latter are much cooler, red giants.

Differences in Stellar Spectra

In a detailed study of variable stars the light curve provides the best means of assigning a star to any particular class of variable but much valuable information may be obtained from the spectrum and at this point it will be useful to discuss the spectral differences among stars and their probable causes.

A spectrum is produced either by passing light through a prism or by the use of a diffraction grating which is simply a bright reflecting surface of glass or metal on which has been ruled many thousands of lines to the inch. In many respects a grating is superior to a prism. The longer wavelengths at the red end are spread to the same extent as the shorter wavelengths at the violet end and also the colours may be more widely spread apart in a grating spectrum.

In 1814, Fraunhöfer investigated the spectrum of the sun, measuring the positions of the dark lines which are present although he did not know the origin of these lines. This did not come until later when Bunsen and Kirchhoff showed that every chemical element produces its own series of lines in the spectrum, the wavelengths of which have all been very accurately measured. We are thus able to tell from the spectrum not only what chemical elements are present in the stars but also the physical conditions of these elements in their widely different environments. For example, an atom of say oxygen or nitrogen may exist in a neutral state or it may have lost a certain number of its outer electrons in which case it is said to be ionised, possessing a positive electrical charge. Not only this, but we are able to tell, quite accurately, the temperature and pressure of the gas which is the origin of the spectral lines. Three main classes of spectra have been recognised and these are often termed Kirchhoff's laws of spectral analysis.

A *continuous* spectrum is produced by an incandescent solid or liquid and, under certain circumstances, by a glowing gas. This is the familiar band of rainbow colours all blending imperceptibly into each other and showing no evidence of bright or dark lines.

An *emission* spectrum, consisting of bright lines against a dark background is the kind normally produced by an incandescent gas, the position of the lines depending upon the chemical constitution of the light source.

An *absorption* spectrum on the other hand is that which is produced by the vast majority of stars, whether variable or non-variable and is made up of a series of dark lines on a continuous background. Such a spectrum is obtained when light from an incandescent solid, liquid or gas shines through gases at a lower temperature. The position of the dark lines corresponds exactly to those of the emission lines which the cooler gases would produce if they were acting as the source of the radiation and are due to absorption of the light of these particular wavelengths from the continuous spectrum.

It is a well known fact that both temperature and pressure alter the lines in a spectrum and when applied to stellar spectra it is important to know which of these two effects predominates. Recent work has indicated that changes in temperature account for most of the differences which have been observed. One must also take into account the absorption produced by gases in the terrestrial atmosphere. Ozone in the upper layers absorbs strongly in the ultraviolet region and both carbon dioxide and water vapour which are present in fairly large amounts cut out much of the infra-red end of the spectrum but in spite of this, it is possible to divide stellar spectra into a number of reasonably well-defined classes.

The initial work was carried out by Secchi between 1866 and 1869 who distinguished four main classes: Type I – hot, white stars; Type II – cooler, yellow stars and Types III and IV which are red stars but showing certain recognisable differences which justified their division into two types.

The system at present used by astronomers is known as the Harvard classification of stellar spectra and is a continuous series with the lines changing smoothly from the hottest stars to the coolest. The sequence was originally in alphabetical order but inevitable changes have resulted in it now being designated as O, B, A, F, G, K, M, R, N and S, their major characteristics being as follows:

CLASS O stars are blue-white in colour and apart from a

very small group of objects known as Wolf-Rayet stars, have the highest surface temperatures of all the stars, up to about 52,000°K. At this extremely high temperature the spectral lines of any metals which may be present in the star's atmosphere will appear in the ultra-violet region and hence are absorbed by the ozone in the Earth's atmosphere and so cannot be easily detected. As in all other classes of stellar spectra hydrogen lines are present in this class but the predominant lines are those due to ionised helium and to oxygen and nitrogen which have been doubly or even triply ionised. Few, if any, variable stars of this class are known apart from some eclipsing systems.

CLASS B stars have a surface temperature around 25,000°K and here the helium lines are those of the element in its neutral state and we find only singly ionised oxygen and nitrogen. Large numbers of these stars are found in the constellation of Orion and for this reason are known as Orion or Helium stars.

CLASS A or Hydrogen stars have a lower temperature about 10,000°K with very intense Balmer hydrogen lines in their spectra. The lines of ionised metals are found although the surface temperature is still too high for the absorption lines of neutral metals to make an appearance.

CLASS F stars are not as numerous as those of the preceding class but they do include most of the eclipsing binaries and a large proportion of the RR Lyrae type variables. The H and K lines of ionised calcium are very prominent, the lines due to hydrogen less intense, with a corresponding increase in the strength of the metallic lines. The surface temperature is in the region of 7,500°K.

CLASS G stars include the sun with a surface temperature of the order of 6,000°K and are yellow stars. Here the H and K lines of ionised calcium are stronger than in Class F while the hydrogen lines are still weaker. These spectra are very characteristic being crossed by many fine absorption lines due to neutral iron in particular and many other metals.

CLASS K stars have a surface temperature of only around 4,000°K and are orange-yellow in colour with Arcturus the brightest example. These are the most numerous type of stars and are found to predominate in the lower galactic latitudes. The lines which are due to hydrogen have been considerably weakened and as a result of this lower temperature we find the characteristic bands due to hydrocarbons. These compounds are those which are relatively stable to heat and can therefore remain undegraded. The major compound to be found is titanium oxide.

CLASS M stars include the large majority of the long period variables. Having a temperature of only about 3,000°K they show the fluted bands of titanium oxide very strongly and since this compound absorbs strongly at the blue end of the visible spectrum, they are all red stars.

CLASS R stars visually resemble Class N mentioned below but are different photographically, the blue and violet portions of the spectrum being somewhat brighter and more pronounced. They are not quite as red in colour as the M or N stars although they do have a slightly lower surface temperature generally around 2,300°K. Unlike the Class M stars, their spectra show characteristic bands due to the presence of carbon compounds.

CLASS N stars are very red and cool possessing peculiar band spectra similar to those of comets and due again to the presence of carbon compounds. They are among the reddest of all the stars with a temperature of about 2,600°K.

CLASS S are also red stars and are nearly all long period variables. They have extremely complicated spectra, somewhat similar to the Class M stars but with broad absorption bands due to zirconium oxide instead of titanium oxide. In addition, bright hydrogen lines and various absorption and emission lines are found. This class was previously included in Class N and was not differentiated from it until 1922. Their temperatures are similar to those of the Class M stars, namely about 2,700°K.

Although not generally included in the above sequence, one further class of stellar spectra must be mentioned in any book on variable stars, namely Class Q which is used for novae. As would be expected, the spectrum of a nova varies considerably after maximum and during the subsequent decline in brightness. Usually bright lines of hydrogen are present although in the final stages these are often absent and the spectrum then resembles a Class O star with very weak lines of singly ionised helium and multiply ionised oxygen and nitrogen.

Since there is a smooth gradation from one spectral class to the next, each may be subdivided further as shown by a suffix, generally a number from 0 to 9 or by a small Roman letter. Here it should be mentioned that certain letters have a particular significance. The letter 'e' is reserved to indicate the presence in the spectrum of bright emission lines. The letter 'k' is used for certain B-type stars which exhibit one or two lines of ionised calcium in their spectra, the origin of these lines lying not in the reversing layers of the stars themselves but due to the presence of interstellar calcium existing between us and the stars. To indicate that the spectral lines are unusually wide or diffuse 'n' is employed, while 's' shows that the lines are sharp. Other peculiarities are indicated by the suffix 'p'.

As well as the above, certain prefixes are also employed, for example 'c' shows that all lines are narrow and sharp, 'g' is used for giant stars, the enhanced lines being fairly strong together with those of hydrogen while low temperature lines are weak. Conversely, the prefix 'd' is used for dwarf stars. Here the enhanced lines are weak with some calcium and titanium lines strong.

It may be seen therefore that the spectrum of a variable star tells us much about the changes which are occurring in the atmosphere of such a body during the course of the light variations and this is extremely important in the study of both extrinsic variables, e.g. those in which no physical changes take place within the star itself such as the eclipsing variables; and the intrinsic variables in which the light fluctuations are due to some mechanism occurring within the star, such as the Cepheids and long period variables.

Three Colour Photometry

The variable star observer frequently makes use of monochromatic filters in the study of these objects, these often being used in conjunction with special photographic emulsions having different sensitivities in the various portions of the spectrum. In the simplest case of all, we may compare the visual and photographic magnitudes of stars. It is well known that the ordinary photographic plate is much less sensitive to red light than to blue and the difference between the two magnitudes is known as the colour index of the star.

With the introduction of isochromatic photographic plates which had been specially sensitised to red and yellow, it was possible to obtain photovisual estimates of brightness in which the spectral sensitivity curve was almost identical to that of the human eye, these estimates not suffering from the marked disadvantage of subjective error.

The technique of three colour photometry has been used extensively in the study of variable stars. In the case of stars having only small amplitudes, a photoelectric device is employed in conjunction with the special filters. The procedure is to measure the star's light variation photovisually, in the blue and in the ultra-violet, to determine their changes in V, B – V, and U – B, the first giving the photovisual light curve which naturally corresponds closely to the ordinary visual one, the second providing us with the blue excess and the last with the ultra-violet flux.

PART ONE

Extrinsic Variables

1. Algol Eclipsing Variables

The ancient astronomers, forced to rely solely on naked eye observations knew of the existence of some of the brighter double stars but were unable to detect any revolution of one component around the other. With the invention of the telescope, many stars previously considered as single were found to be close visual doubles. The term double star is now normally retained only for those which are apparently close together although in reality one is considerably nearer to us than the other. Those stars which are sufficiently close to revolve about a common centre of gravity are better known as *binaries*, or physical doubles.

Many of these binaries are so far apart that the components may be separated in the telescope and take a hundred or more years for one revolution. Others are so close that even in the largest instruments they appear as a single point of light. In this case there are two methods of proving that they are indeed binaries.

If the plane of the orbit of the stars as they move around their common centre of gravity lies on, or very close to, the line of sight as seen from Earth, then the components will mutually eclipse each other during the period of one revolution with a subsequent diminution of light. Such systems are known as eclipsing binaries and form a large and important class of stars known as *extrinsic variables*, a term which simply denotes that no physical change in either star is the main cause of the light fluctuation.

In the second instance where no eclipse occurs, the proof of duplicity is given by the spectroscope, hence such systems are known as spectroscopic binaries. To understand the reason for the changes in the spectrum of such a star we must here consider a fundamental principle of spectroscopy known as the

Doppler Effect which has proved to be extremely useful in the study of variable stars as a whole. If the distance between an observer and a light source is decreasing the lines in the spectrum are shifted towards the violet end of the spectrum as compared with the lines of a stationary source and conversely, if the distance is increasing, the lines are shifted towards the red. In addition, the amount of the shift is directly proportional to the relative velocity of recession or approach. Quite often the stationary spectrum is formed in the observatory by striking an arc between iron electrodes and photographing this both above and below the spectrum of the star when a direct comparison may be made. Since all stars of spectral class F and later contain numerous lines due to iron in their spectra this method may be used for the great majority of stars. In other cases, the spectrum of hydrogen has proved useful. It must be emphasised that only motion directly towards or away from the observer can be detected in this manner. Movement in any other random direction cannot be measured by means of the Doppler Effect.

Providing that the star is not too faint it is a relatively simple matter to photograph the spectrum of such a binary system and we find that when one star is directly behind the other the common lines in their spectra will coincide since the stars are then moving at right angles to the line of sight and the Doppler shift is zero. However, when they are side by side, the lines due to the component moving towards us will be shifted towards the violet while those of the star moving away from us will be correspondingly shifted to the red resulting in a separation, or doubling, of each set of lines. They will therefore be seen alternately single and double proving that the components are in motion about each other.

Spectroscopic binaries as such are not generally classed as variable stars since there is no detectable change in their light, only their spectra indicating that they are double star systems and they will not be dealt with further in this book. Only those stars in which eclipses occur will be considered here.

As mentioned earlier, Algol in the constellation of Perseus was possibly the first variable star to be discovered, its name being given to it by the early Arab astronomers. At regular intervals of 2 days 20 hours and 49 minutes its light fades from magnitude 2.20 to 3.47, the primary eclipse lasting for almost two hours. The system has been shown to consist of a bright star of spectral class B8 and a much fainter companion, the difference in brightness of the two being such that special photoelectric equipment is necessary for the accurate measurement of the secondary eclipse found midway between the principal minima and due to the brighter star passing in front of the fainter. When this happens, the light fades by only 0.06

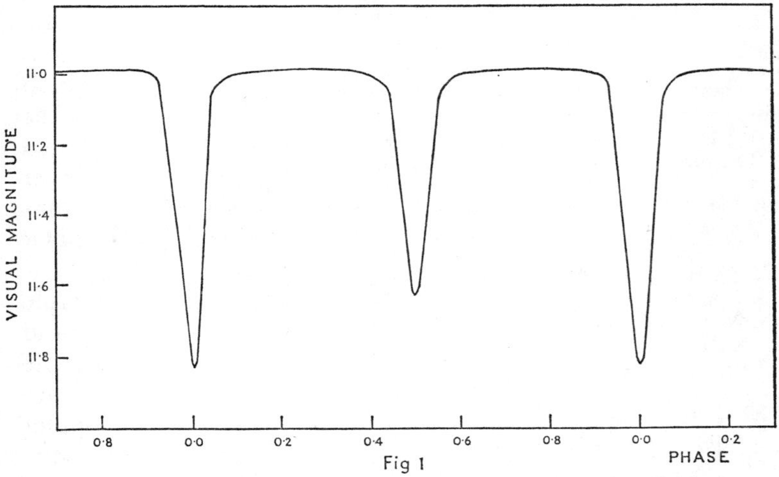

Fig. 1 – Light curve of RW Lacertae, a typical Algol variable.

of a magnitude. The spectroscope shows quite conclusively that both primary and secondary minima occur when the Doppler shift is zero; that is when the stars are behind each other.

Algol is the prototype of almost five hundred similar known eclipsing variables all characterised by having almost flat maxima and well-defined minima (Fig. 1), the component stars being sufficiently well separated for them to be essentially spherical in shape. We shall see in the next chapter that there is another large group of eclipsing variables where this is not the case and the stars are so close together that they are virtually in contact and are distorted by gravitational and tidal forces into marked egg-shapes.

Algol itself is bright enough for the whole of its light cycle to be followed with the naked eye and several more of these variables are within easy reach of binoculars. The great importance of these variables lies in the fact that we can learn far more about the nature of the individual stars than would be possible if they were single.

Table 1 gives a list of some of the brighter Algol variables:

Table 1

ALGOL VARIABLES

Star	Maximum	Primary Minimum	Secondary Minimum	Period (days)	Spectrum
Algol	2.20	3.47	2.26	2.86731016	B8
RY Aqr	8.80	10.11	8.91	1.966598	A3
R Ara	6.05*	6.97	6.20	4.42509	B9
β Aur	2.07**	2.16	2.16	3.960077	A2 + A2
U Cep	6.74*	9.81	6.77	2.4929005	B8 + gG2
VV Cep	6.62*	7.42	6.63	7430	cM2ep + B0
Y Cyg	7.00	7.64	6.63	2.9963331	O9 + O9
Z Her	7.20	7.91	7.42	3.992795	F2 + F2
TX Leo	5.70**	5.80	5.72	2.445	A0
δ Lib	4.79*	5.89	4.88	2.32734906	A1s
RR Lyn	5.60**	5.97	5.84	9.9450	A6s
U Oph	5.80**	6.50	6.40	1.677346	B5n + B5n
EE Peg	7.01	7.54	7.12	5.256	A0
AG Per	6.49**	6.80	6.74	2.028729	B3 + B3
U Sag	6.36*	9.04	6.46	3.380618	B9 + G2
RS Sgr	6.00*	6.97	6.22	2.4157	B5
V356 Sgr	6.81*	7.92	7.20	8.8965	B8 + A1
V505 Sgr	6.50	7.31	6.54	1.182871	A2
μ^1 Sco	3.00*	3.31	3.20	1.6642683	B3 + B6
RZ Sct	7.70*	8.87	7.75	15.19016	B2
λ Tau	3.50**	4.00	3.59	3.95295	B3
TX UMa	6.83**	8.92	6.88	3.0637175	B9 + gF2
α Vir	1.21**	1.30	1.28	4.014160	B3n
Z Vul	6.97	8.60	7.11	2.454926	B3n + A3
RS Vul	6.90	7.63	7.00	4.477666	B5 + A2

* Photographic magnitudes

** Photoelectric magnitudes

The case of VV Cephei should be particularly noticed for its exceptionally long period of 7,430 days. This star is one of a small group of eclipsing variables having long periods and these will be discussed in more detail later in the chapter. In the meantime, we shall look at some of the deductions which may be made simply from the light curve itself.

Elements of the Orbit

A tremendous amount of data has been accumulated relating to the light curves of these variables and remembering that each component is directly influenced by the gravitational field of the other, we have here systems somewhat analogous to the solar system, differing only in that there are only two bodies in general to be taken into consideration and both are far more comparable in size and mass.

If the orbit is circular as is the case with many close binaries, the stars will move with a constant velocity and consequently in the light curve, the primary and secondary minima will be equally spaced. On the other hand, if the orbit is elliptical, this will not be so. Kepler's second law of planetary motion states that the line joining the sun and any planet will sweep out equal areas in equal times. Quite simply this means that where the orbit is eccentric these equal areas will include different lengths of the planetary path and consequently the planet will move more quickly when it is near to the sun than when it is further away.

Similarly in the case of a binary system the stars will move with varying speeds around their common centre of gravity and this will result in the minima being unequally spaced. Thus an idea of the shape of the orbit may be derived from the relative positions of the minima as shown in Fig. 2.

Most of the Algol variables possess orbits which are approximately circular although some are known which exhibit a high eccentricity and especially in the latter, a secondary effect has been discovered recently which is known as apsidal motion and has the effect of actually shifting the relative positions of the minima in the light curve. However, discussion of this will be postponed until later in the chapter.

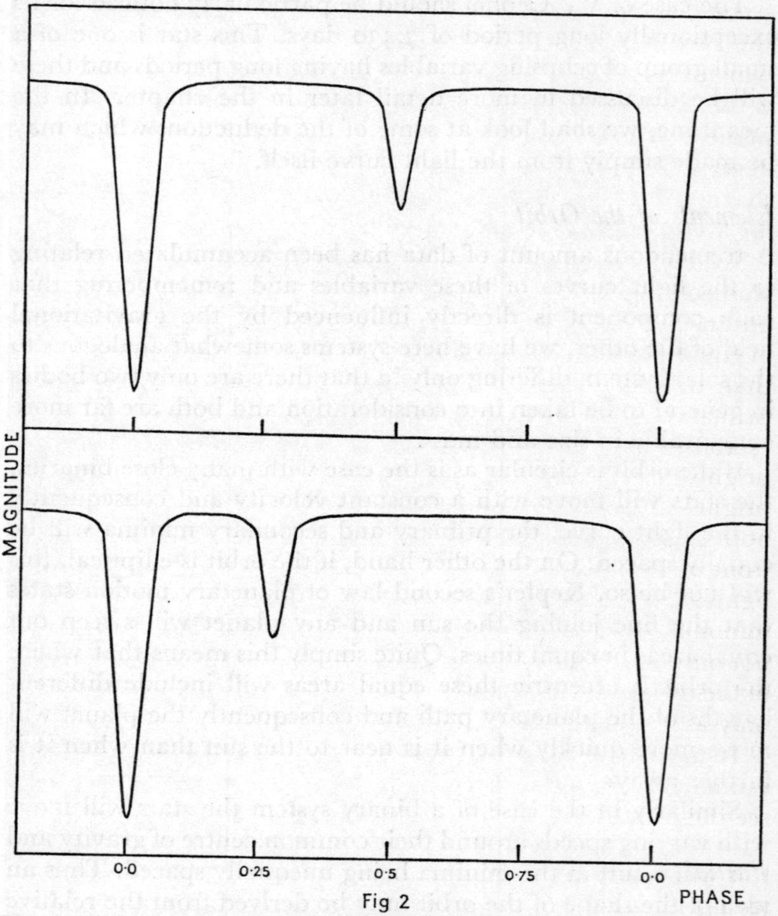

Fig. 2 – Light curves of Algol variables showing the variation in the position of the secondary minimum due to the ellipticity of the orbit.

Relative Brightness of the Components

The depth and shape of the minima themselves also tell us something of the relative brightness of each star. In the simplest case where we have two components of approximately equal size and of the same luminosity (a condition which is only rarely encountered in this type of variable), both primary and secondary minima will be of equal depth since the total light of the system will be reduced to the same extent irrespective of which star eclipses the other. As will be seen from Table 1 this is the case for β Aurigae and Y Cygni. Also, assuming that the plane of the orbit is 90° to the sky, that is edge-on to the observer, the minimum of the light curve will dip to a fairly sharp point since the actual moment of totality will last only for a very short time compared with the total period.

In the majority of eclipsing variables which have been studied there are several deviations from the simple scheme outlined above. For example, the two minima are usually not of equal depth indicating that one component is much brighter than the other and by careful measurements of the depths at primary and secondary eclipse the relative brightnesses may be established. Thus if we assume that one star is only a quarter as bright as its companion then when the faint one eclipses the bright star, we will receive only a quarter as much light as when both stars are side by side; and when the bright star passes in front of the faint companion, we get three quarters of the total light. The ratio of the depths at primary and secondary minimum will therefore be 3:1. Thus the light curve gives us not only the shape of the orbit but also the relative brightnesses of the two components even though in the telescope they appear only as a single point of light.

It must be emphasised however that this gives only a true estimate when totality occurs at each eclipse. As we shall now see the situation as we find it is seldom as simple as this.

Partial Eclipses

By far the greater number of Algol variables consist of stars of unequal size and because of this their light curves exhibit certain peculiarities which, although making the interpretation somewhat more difficult, provide us with additional information concerning these interesting binary systems.

Where one component is smaller than the other only a partial eclipse can occur whenever this smaller star passes in front of the larger and as a result the secondary minimum appears flat rather than pointed. A partial eclipse almost invariably occurs at secondary minimum since it has been found that in most, if not all of the Algol systems, it is the brighter star which is the smaller of the two. When considering the possible evolution of these stars we shall see that this is a somewhat surprising fact which leads us to postulate a rather singular process by which these stars may have been formed.

When we consider the effect of inequality of size on the shape of the primary minimum we find that here too there is a slight flattening at the base of the minimum due, in this case, to the fact that the period of totality lasts longer than it would do if the stars were of equal size.

There are a few Algol variables, notably ζ Aurigae, in which the difference in the sizes of the components is so great that whenever the smaller, brighter star crosses the disc of the huge, supergiant companion it should really be called a transit rather than an eclipse. This particular star has the remarkably long period of 972 days and consists of an orange supergiant of spectral class K5 having a diameter of almost 180,000,000 miles and a much smaller, brighter star of class B9 only three times larger than the sun. The secondary minimum lasts for thirty-eight days and a very sensitive photometer must be employed to detect the extremely slight fading of light during the transit.

This star also affords us an almost unique opportunity to study the atmosphere of a star independent of the star itself. The atmosphere of the K5 supergiant is so deep that every 972 days there is a period of two or three weeks just before and after primary eclipse when the light of the bright star actually shines through the fainter companion's atmosphere with subsequent changes in the B9 spectrum. These changes enable us to examine both the chemical composition and the thickness of the atmosphere of the supergiant. One surprising result that has emerged from such studies is that the depth of the chromosphere is not the same at different eclipses. Obviously, considering the length of time during which the small star shines through the large component's atmosphere, the chromosphere of this star extends almost to the very centre of this gigantic sun. This is, of course, exactly what we would expect. Observationally, it has been discovered that whereas the sizes of the stars vary enormously, the range of masses is relatively restricted.

Consequently their densities must also vary over a tremendously wide range and in the case of these supergiants the density must be extremely low, even less than that of a good vacuum here on Earth. Such systems are, unfortunately, not numerous and a possible reason for this rarity will be given when we discuss the evolution of these curious binary systems.

Apart from an inequality of the sizes of the component stars there is a further cause of partial eclipses. The inclination of the plane of the orbit may deviate appreciably from 90° so that only a part of one star covers the edge of the other. This is actually the case with Algol, the secondary minimum as seen from Fig. 1 being rounded rather than pointed. Naturally, since these binaries are aligned quite randomly in space, the inclinations of their orbits are also completely random.

Reflection Effect

A further noticeable aspect of the light curves of many Algol variables is that the maximum is not truly flat but shows a slow rise from the end of primary minimum and a correspondingly slow dip before secondary minimum. The reason for this is that both hemispheres of the fainter star are not of the same brightness. When the side facing the bright companion is turned towards us this star gives more light than when the opposite face is turned in our direction. This is known as the Reflection Effect although actual reflection of light from the star's surface is only partly responsible for the light changes. It is also probable that the intense radiation from the hotter companion causes excitation of the atoms in the faint star's outer layers resulting in a higher temperature and increased light.

From the foregoing account it will thus be seen that even the light curve alone gives us valuable information concerning the two stars which make up an Algol system. When taken in conjunction with the spectrum of the variable, even more information may be obtained.

Radial Velocity

Where the spectrum of the bright star can be differentiated from that of the fainter, its Doppler shift can of course be measured giving its velocity in its orbit. This is normally measured in kilometres per second and as the time of revolution is known from the period, the circumference and hence the diameter of the orbit may be calculated. Where the orbit is not circular, the eccentricity provides us with the means of determining the lengths of the semi-major and semi-minor axes. The time of the eclipse as measured from the two minima, coupled with the orbital velocity, will then give the diameters of the two stars in kilometres.

By plotting the measured velocity against phase it is possible to obtain a radial velocity curve, from the symmetry of which the eccentricity of the orbit may be derived with a high degree of accuracy. Where the radial velocity curve is symmetrical the orbit is circular, any departure from symmetry indicating an elliptical orbit. Recent studies however have shown that the position is not quite as simple as was at first thought.

The variable U Cephei has been known for more than a century and was early recognised as an Algol type variable. The primary minimum lasts for 10 hours and occurs every 2.4929 days. Accurate measurements have shown that the period of this star is not constant but has been slowly increasing and in addition there are irregularities whose cause is not yet shown.

All of the early radial velocity curves of U Cephei were highly unsymmetrical indicating an eccentricity of between 0.3 and 0.5 whereas photoelectric observations showed a very shallow minimum which occurs exactly halfway between the primary minima showing that the stars apparently move in a circular orbit. Here therefore was a case in which the light curve and radial velocity curve were completely at variance.

The clue to the reason for this curious anomaly came when it was found that in the spectrum of U Cephei the broad absorption lines of hydrogen were also unsymmetrical. Struve found this asymmetry to be due to the presence of a very narrow absorption line superimposed on a Stark broadened line. The Stark Effect is an optical phenomenon due to electrical fields within the atmospheres of stars and shows itself as a splitting of the spectral lines (in the case of low dispersion spectra only a broadening of the lines appears). Although there are no overall electrical fields operating within the stars themselves, in contrast to magnetic fields which are present on a large scale, at high temperatures, a large proportion of the atoms are ionised being then no longer neutral but electrically charged, leading to the creation of numerous small-scale electrical fields which can be transferred from one atom to another in the course of their random motion.

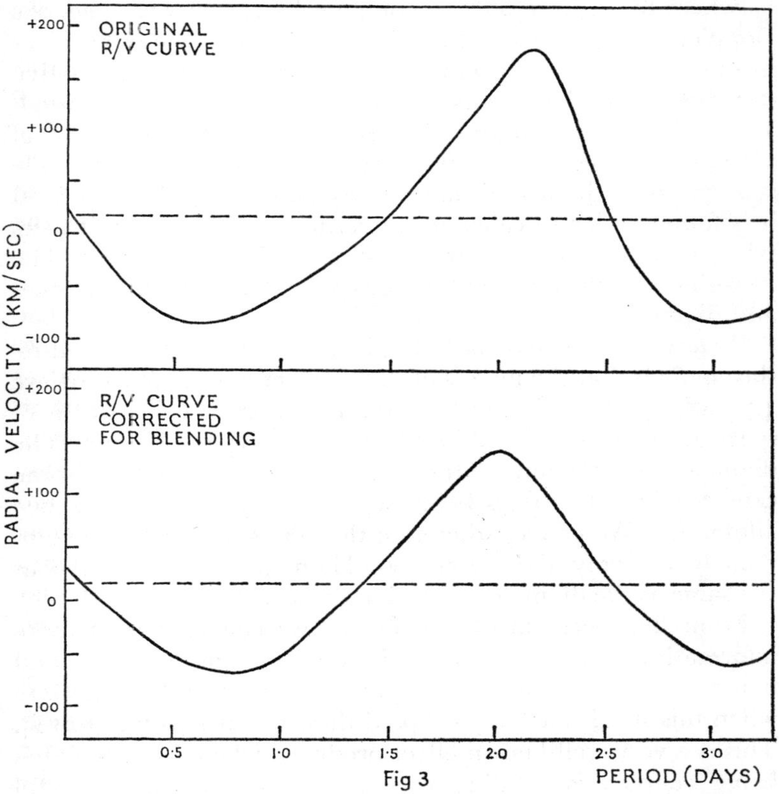

Fig. 3 – Original and corrected radial velocity curves for U Cephei.

Hardie has demonstrated that in the spectrum of U Cephei, the positions of the narrow absorption lines vary as the stars move in their orbit being displaced towards the violet at maximum velocity of approach and towards the red at maximum velocity of recession. When the radial measurements have been corrected for this effect, the lines then give an almost symmetrical radial velocity curve which is in quite good agreement with the light curve. (Fig. 3.)

Mass and Diameter

In those cases where the orbit has been shown to be circular, or where the spectra of both components can be observed, the Doppler shift of one set of lines compared with the other gives us the relative velocities of the stars and from this their relative distances from the common centre of gravity can be found. With this information and using Kepler's and Newton's laws of planetary motion (which, of course, are also obeyed by double star systems), the actual masses may be computed. In general it is found easier to calculate these in terms of the sun's mass taken as unity rather than to estimate the true masses. The densities of each star may then be calculated from their masses and diameters.

Reverting to the case of U Cephei, it has been established by this method that the masses of the brighter and fainter stars are 4·7 and 1.9 times that of the sun. The more massive star is of spectral type B8, the other being a G-type subgiant. The diameters are 5.8 and 9.2 times that of the sun respectively, the centres of the two stars being separated by about 10,000,000 kilometres. We can see therefore that the surfaces of the components are only about 4,500,000 kilometres apart, just about the same as the diameter of the larger star.

From the spectrum of the B8 star we can gain even more information. The absorption lines of helium, calcium and magnesium are found to be extremely broad and diffuse and when this star is partially eclipsed they are curiously distorted. This is a very well-known effect produced when a star which is being eclipsed is rotating very rapidly on its axis. Careful measurements give a figure of almost 300 kilometres per second for the rotational velocity of this star.

Synchronisation of Orbital and Rotational Velocities

By a comparison of the period of rotation with the orbital period, we find that the B8 star is rotating five times faster than would be the case if rotation and revolution were in synchronisation. When we come to examine the spectrum of the G-type subgiant, we find the opposite state of affairs. Here the absorption lines are all extremely narrow indicating a much slower rate of axial rotation. It is not possible to obtain a result as accurate as that for the B8 star since the G-type spectrum can be seen only during secondary eclipse but it has been estimated that the less massive star rotates only about half as fast as is necessary for synchronisation.

The investigations of the last twenty or thirty years have established the existence of several Algol variables in which rotational and orbital periods are not synchronised. In every one of them it is the smaller, brighter and more massive component which rotates too rapidly. We must therefore now consider the cause of non-synchronisation in these variables, remembering that in the majority of cases studied this appears to be the case.

One thing is known, namely that where there is a flow of gas from the larger star to the smaller, synchronisation does not occur. The suggestion has been made that when initially formed, these binaries were all synchronised but that gradually a belt of gas was formed which carries off both mass and angular momentum from the larger, less dense star. Such a connecting gas stream has the inevitable effect of slowing the rotation of the large star and speeding up that of the hot primary. The presence of a luminous ring of hydrogen gas has been proved spectroscopically as described below for the variable RW Tauri but one important question still remains to be answered. Does the entire B8 star rotate with this very high velocity, or only the outermost layers?

At the present time it is not possible to answer this question observationally. All that we can say is that since the fast rotation is measured by the Stark-broadened lines of hydrogen, the rapid rotation certainly occurs as deep as the reversing layers of the star where these broad absorption lines originate.

Luminous Gas Streams

The variable RW Tauri is normally of photographic magnitude 8.00 but every 2.768 days it fades to 12.27 magnitude, the eclipse, which is total, lasting for 84 minutes. The brighter component of this system is a white star, spectral class B9e, while the other is a yellow Ko star having about twice the diameter of the primary. The bright hydrogen emission lines in the spectrum of the B9e star were found by Joy to behave in an extremely puzzling manner. As the eclipse begins, these lines are displaced towards the red end of the spectrum showing that their source is moving away from us. At mid-eclipse they vanish altogether, reappearing again just after eclipse but now shifted by an equal amount towards the violet. Only these emission lines of hydrogen are affected in this curious way, all other lines remaining in their normal positions.

On the basis of these observations, Joy suggested that they originate in a ring of hot hydrogen gas surrounding the brighter star and somewhat smaller in diameter than the larger, fainter companion. The shifts are due to the larger star covering first one side of the ring and then the other during eclipse. From accurate measurements of these shifts it is estimated that this gaseous ring is rotating in about fourteen hours with a velocity of 350 kilometres per second. The diameter of the ring is believed to be about four times that of the sun, namely about 3,540,000 miles.

Limb Darkening

It has been recognised for perhaps many centuries that when the glare of the sun is sufficiently reduced so that the disc may be viewed without discomfort, the disc or photosphere is noticeably darker around the edges than in the centre. The light emerging from different levels in the solar atmosphere is strongly scattered by the overlying layers of gas, the degree of scattering being determined both by the path length which the light has to travel and the wavelength of the light itself. The shorter the wavelength, the more strongly the light is scattered. A result of this is that the outermost layers possess a certain opacity which means we can see only a certain distance into the atmosphere. Because of this, the light reaching us from the centre of the disc originates from the deeper, hotter levels compared with that emitted from the limb. As might be expected, since blue light has a shorter wavelength than red, the effect is more pronounced when viewed in blue light than in red or yellow.

Theoretically, we must presume that this holds true for most, if not all stars and should be detectable in the case of eclipsing binaries, especially if they are viewed in blue light. The procedure is to make use of three colour photometry and compare the light curves in red and blue light. Where limb darkening is marked (and this is not always the case since with some stars it is so small as to be virtually undetectable), the descending and ascending branches of the minima will appear steeper and shifted slightly. A great deal of work has been carried out in the past two or three decades on this seemingly trivial phenomenon.

VV Cephei

As shown in Table 1 this variable has the exceptionally long period of 7,430 days and is one of a small group of Algol variables having these remarkably long periods. One of these, ζ Aurigae, has been mentioned earlier. Like this star, VV Cephei consists of two components which differ in size and spectral type, representing the two extremes in this respect. The masses have been estimated as 30 and 50 times that of the sun with spectral classes B9 and M3 respectively. Several models of this curious eclipsing system have been devised by various astronomers; the latest by Margherita Hack (1961) is of a large M-type star almost twice as massive as the smaller, very hot star which is totally embedded within a shell of semi-transparent gases which are the cause of the eclipse. The diameter of this shell is considered to be many times that of the cool supergiant which itself is probably more than 1,000 times as large as the sun. Here, therefore, we have two components on a truly large scale illustrating the tremendous diversity which exists among the stars.

Apsidal Rotation

Although many Algol variables show no change over a long period in the intervals occurring between eclipses, several are known in which there is a periodic shift of the secondary minimum with respect to the primary, a circumstance which is of particular importance in the study of these variables since it allows us to obtain more information concerning the structure of the interior of a star, information which would not otherwise be available to us. We have already seen that if the orbit of a binary system is highly eccentric the two minima are not equally spaced in the light curve but where this is the only parameter with which we are dealing, this inequality will remain constant. Where this inequality shows periodic changes it is obvious that some other force is operating within the system and about forty years ago it was recognised that some binary systems exist in which the orbit itself progresses over a complete revolution of its major axis. The latter is usually termed the apse line or the line of the apsides, the effect itself being known as the apsidal rotation of the orbit.

The last forty years of theoretical investigation into this periodic motion which turns the orientation of the orbit in space through 2π radians in an apsidal period which is always long compared with the primary period of the variable itself, has established the fact that apsidal rotation will occur with any elliptical orbit whenever the common centre of gravity does not act like a point source; in other words the gravitational field of the system does not obey the inverse square law as put forward by Newton.

There are three main reasons why this may not be the case. If there is a third body close enough to the system for its gravitational field to produce an appreciable effect on the pair then apsidal rotation of the binary will take place (the apsidal rotation of the moon under the outside influence of the sun is a well-known example of this). Any distortion of one of the components from a truly spherical shape will also produce this effect and lastly it may be that one of the constants used in the equations to describe the orbital motion may to a certain extent be variable. The slow change in the perihelion of Mercury which provided a striking confirmation of the General Theory of Relativity is quite a good example of this form of perturbation.

Whichever of the above forces may be operative in the case of apsidal rotation in any particular binary system, one of the most important consequences is that the elliptical orbit is not completely closed. (Fig. 4.)

As may be expected from a consideration of Fig. 4 the period of apsidal rotation will be extremely long, possibly amounting to thousands of the primary periods even in the most favourable cases and there are several other difficulties in the accurate measurement of these periods. Primarily, of course, the apse migration can only be detected in the case of eclipsing binaries and accordingly we are restricted to an investigation of these variables.

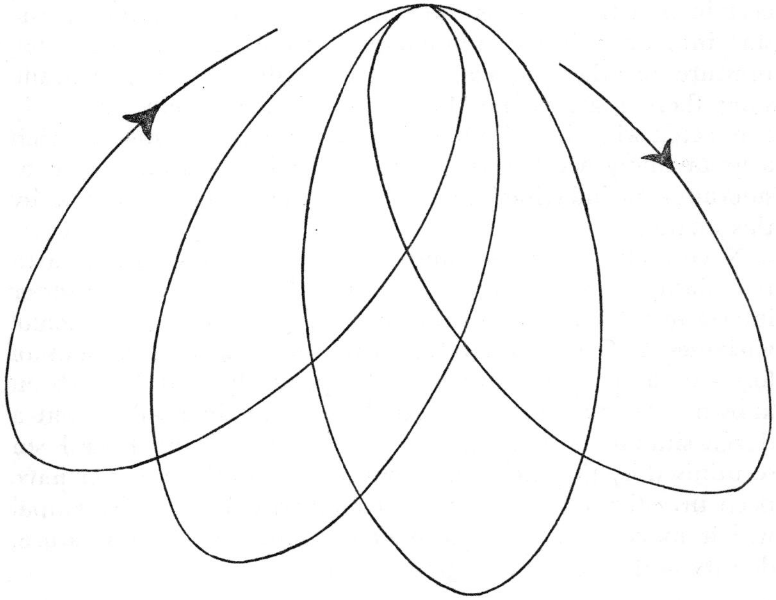

Fig 4

Fig. 4 – Open elliptical orbit of a close binary system illustrating apsidal rotation.

Owing to the fact that the displacement of the secondary minimum with respect to the primary depends on the eccentricity of the orbit and the varying velocities of the stars at periastron and apastron, the smaller the eccentricity, i.e. the more nearly circular the orbit, the more difficult it will be to measure the magnitude of this effect. However, since a high eccentricity is normally associated with a long period, this further restricts the number of stars suitable for such an investigation.

Another problem which reduces the number of suitable candidates still further is that of determining the type of perturbation which is responsible for the rotation of the apsides. If we are to utilise this motion for obtaining indirect knowledge of the internal structure of the stars, we have to be absolutely certain that the cause is the simple one of deformation of the star into an ellipsoid, implying a gradation of composition, pressure, or other physical properties within the interior of the star; there must be no third body in the system physically connected with the eclipsing binary. Multiple systems, of which a great many are known, must be eliminated from any consideration of inferring the physics of the interior of a star by this method.

Nevertheless, a small number of variables are known with reasonably short periods of apse rotation and with no other interfering bodies to complicate the problem. The Algol variables Y Cygni and GL Carinae have apsidal periods of forty-six and twenty-five years respectively and have been known for some years now. At the present time only about a dozen similar variables have been discovered and have been scrutinised by this method. About four times this number have been investigated by an indirect method developed by Kopal which makes use of a parameter relating the temperature, density and pressure in a gaseous sphere.

The Evolution of the Algol Variables

Finally, we come to the evolution of these variables. One special feature of the Algol binaries is that the larger component is the fainter of the two. At first sight this may not appear unusual but if we look a little deeper into the matter, several peculiar facts become apparent. Quite evidently, the two components of these eclipsing systems cannot both be main sequence stars (for a detailed discussion of the main sequence and the Russell-Hertzsprung diagram, see Chapter 5), since if this were so, the fainter star would have to be the smaller.

Two alternatives are therefore open to us. Either the two stars were not formed at the same time, or the evolutionary track of the components of these systems is a highly unusual one.

Considering the first possibility, we know from observation that the bright star lies on or very close to the main sequence and consequently, for it to be the larger of the two the faint companion must have evolved to the right of the Russell-Hertzsprung diagram. This would certainly result in it becoming larger and fainter. Such a picture does however introduce the added complication that the stars must be of different ages, that the bright star must be younger than its companion. If both stars were initially main sequence stars of approximately the same age then it is difficult to see how one component could begin to evolve without the other doing so, particularly since they are very close together.

From our present knowledge of the formation of double – or multiple – star systems, it seems certain that such stars are all formed at about the same time from a large gas cloud; certainly there can be no wide difference in their ages such as we have just postulated. We are therefore forced to the conclusion that some other evolutionary process is responsible for this curious anomaly.

Hoyle has suggested an evolutionary mechanism based on the interchange of mass between the stars which fits the observed facts very well. If we assume that the two stars of an Algol binary are initially on the main sequence but one star is much brighter than the other, then the bright star, being further up the main sequence, will be the first to begin evolving to the right. As we would expect, this star expands as it evolves but since the two components of the system are so close together the expanding star literally engulfs its companion.

The stage is now set for the star which has been enveloped to begin acreting hydrogen from the vastly extended outer envelope of the giant star. Incredible as it may seem, this is quite possibly what actually happens and in taking material from the large star, the total amount of hydrogen in the latter is effectively reduced relative to the helium in the core. This has the effect of speeding up the evolution of the distended star to the point were contraction occurs so that it returns towards the main sequence but since much of its hydrogen has been lost the resulting increase in brightness is far less than would normally have been the case. In the meantime, the other star, having gained mass in the process, moves up the main sequence. The end result will be that which we actually observe in the majority of Algol binaries.

We must however also consider the case where the two stars are not close enough for one star to engulf the other. When this is so, it is then possible for the brighter of the two to begin evolving while the fainter star remains on the main sequence. A point will then be reached when we have a system made up of a red star of very large size with a high luminosity and a smaller star with a brightness perhaps 100 to 200 times that of the sun. These are indeed the Algol variables such as ζ Aurigae and VV Cephei already mentioned having very long periods. We thus see that even in this single class of variable stars, almost all stellar types are represented covering an extremely wide range of sizes, temperatures and masses.

2. β Lyrae Eclipsing Variables

In our study of the Algol variables we saw that the components are generally sufficiently far apart for them to be approximately spherical in shape; in other words the gravitational force acting upon each is not strong enough to distort them to any great extent. We must now consider a second large group of eclipsing variables in which the stars are so close as to be virtually in contact. Here the force of gravity is so tremendously high that the tidal effect on each component exerted by the other pulls them out into an elongated egg-shape.

The prototype of this class of variable is the beautiful third magnitude star β Lyrae which varies between 3.4 and 4.3 magnitude in a period of almost thirteen days. Since the stars are not even approximately spherical the light curve shows a continuous variation even when no eclipse is taking place. Fig.

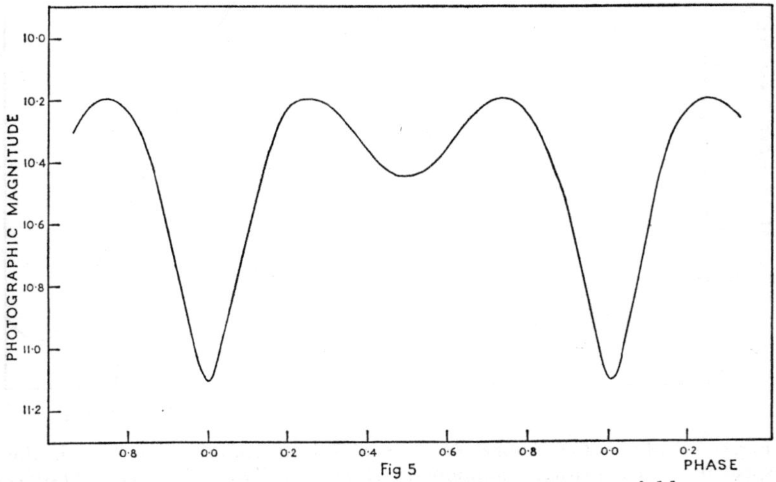

Fig. 5 – Light curve of U Scuti, a typical β Lyrae variable.

5 shows the light curve of a typical β Lyrae variable obtained by the author from 268 visual observations.

Owing to the inevitable braking effect of tidal friction, the periods of rotation and revolution of these stars are almost equal, i.e. they are in synchronisation so that at eclipse we see the pair end-on while midway between eclipses the elongated sides of both stars are presented to us. At other times, a varying area of surface is exposed and as a result there is not even an approximately flat region in the light curve at maximum which is so characteristic of the Algol variables.

In the same manner as for the Algol variables, a study of the light curve of a β Lyrae eclipsing system provides us with valuable information about both the shape of the orbit and the relative brightnesses of the two component stars. Generally speaking, the secondary minimum is deeper and more well-defined than that found in the majority of the stars of the previous class and as we might expect from the fact that the orbits are far less eccentric, the radial velocity curves are more symmetrical.

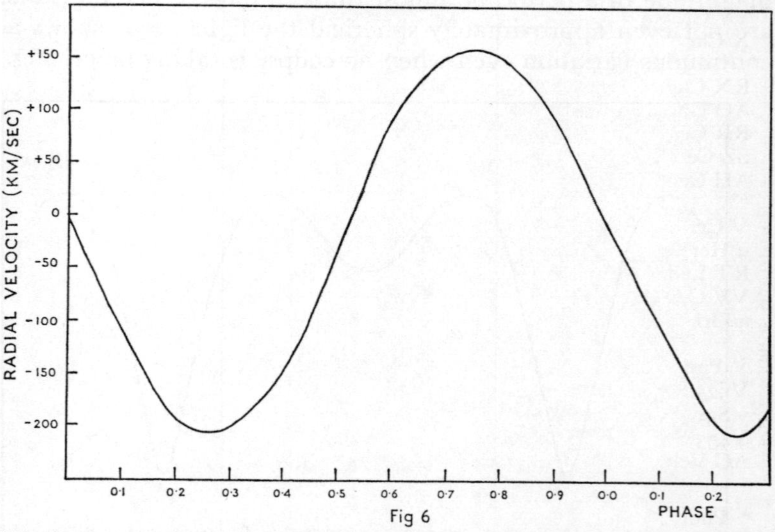

Fig. 6 – Radial velocity curve of β Lyrae derived from the Si II lines in the spectrum of the primary component (*Astrophysical Journal by kind permission of the University of Chicago Press*).

Fig. 6 shows the radial velocity curve for β Lyrae derived from the lines of singly ionised silicon in the spectrum of the primary star.

Much work has recently been done on the β Lyrae system, principally by Struve and his colleagues who have shown that changes in luminosity occur from cycle to cycle, presumably due to changes in the opacity or amount of gas in the system

Table 2

β LYRAE VARIABLES

Star	Maximum	Primary Minimum	Secondary Minimum	Period (days)	Spectrum
β Lyr	3.40	4.31	3.80	12.9080006	B8p
AN And	6.00**	6.16	6.09	3.2196	A7s+A
S Ant	6.39	6.83	6.79	0.6483387	A8+A8
σ Aql	5.00**	5.18	5.14	1.95026	B3+B3
TT Aur	8.06*	9.02	8.43	1.33273	B3+B3
SZ Cam	7.00*	7.27	7.22	2.6984166	Bon
TU Cam	5.20	5.40	5.24	2.93328	A1
UW CMa	4.47	4.81	4.80	4.3934	O8+O8
YY CMi	8.51	9.12	8.90	1.0740651	F5
X Car	8.05*	8.60	8.57	1.0826310	Ao+A1
RX Cas	†10.04*	11.13	10.32	32.3166	gG3+gA5e
AO Gas	5.80**	6.01	6.01	3.52341	O8n+O8n
RR Cen	7.65*	8.11	7.94	0.605686	F2
SZ Cen	8.30	8.92	8.82	4.10796	A2
AH Cep	6.63	6.85	6.79	1.77465	Bo+Bo
V Crt	9.52	10.21	9.90	0.702034	A6
u Her	4.60	5.14	4.80	2.0510277	B3+B3
RT Lac	8.80	9.61	9.52	5.073977	G9+K1
VV Ori	5.10**	5.44	5.26	1.4853787	B2n+B8
η Ori	3.20**	3.55	3.23	7.98922	B1
V Pup	4.53*	5.14	5.01	1.4544867	B1+B3
V525 Sgr	8.52	9.22	8.80	0.705125	A2
ν Sgr	4.26*	4.41	4.36	137.00421	B8p+F2p
V453 Sco	6.56	6.99	6.98	12.00421	Bose
AC Vel	8.53*	9.05	9.04	4.5622426	B8

* Photographic magnitudes

** Photoelectric magnitudes

† The magnitude at maximum varies between 9.82 and 10.28 in a period of 517.6 days.

and in addition there are unexpected deviations in the radial velocity curve over a period of only three years, these taking place near the secondary minimum of the primary star. This result is not easy to explain since genuine changes in the orbital elements would not normally be anticipated in so short a time. Table 2 gives details of some of the brighter β Lyrae variables.

Synchronisation

In the previous chapter on Algol variables it was mentioned that in many cases synchronisation between rotational and orbital periods breaks down, the early-type primary rotating too quickly. There are, however, a few cases known in which the primary rotates too slowly for synchronisation and this is actually the case with β Lyrae, the B8 star having an axial rotation of only 40 kilometres per second whereas the synchronised value would be nearer 200 kilometres per second. Some astronomers believe that in β Lyrae the B8 star is the least massive of the two components and observation indicates the presence of a powerful gas stream carrying matter towards the other star, resulting in a deceleration of the rotational velocity of the B8 companion. Presumably there is a corresponding increase in the axial rotation of the secondary but since its spectrum cannot be defined it is impossible to confirm this observationally.

Recent studies of this eclipsing system during which it was observed simultaneously photoelectrically and spectroscopically show that, in those cycles where the primary minimum in the light curve is abnormally deep, the shell lines of the gas stream are stronger and have unusual displacements. Such observations demonstrate that a part of the light loss at minimum is due to, not only the eclipse, but to absorption by this shell of hydrogen gas. Furthermore, the velocity measurements are also affected by the rotation and density of the gas streams in the system.

From certain irregularities in the radial velocities as first determined, it was considered that there might be an invisible third body in this system but recent redeterminations of the values have failed to provide any convincing evidence of this and such a view is not now generally held by most astronomers.

Masses of β Lyrae Variables

Let us now turn to a consideration of the masses of the components of β Lyrae variables. In the previous chapter we saw how, as in the case of U Cephei a typical Algol variable, the primary star is the more massive of the two. The determination of the masses of the two stars which form a β Lyrae system is by no means as easy or as straight-forward as in the case of the Algol binaries. The mere fact that the periods of many of these stars are increasing indicates that there is a mass loss in these systems, providing further confirmation of some earlier spectroscopic evidence found for β Lyrae itself that mass is streaming outward from the system.

A large number of β Lyrae stars have been studied during the past three decades and from the data at present available it would appear that here, too, the secondary is the least massive; in most cases it is believed to be a subgiant star. There are many difficulties associated with the determination of the orbital elements of these particular variables and it is only recently, with the introduction of computer techniques, that accurate elements have been obtained. Indeed, in some instances, as with V Puppis, results obtained by this method have confirmed that the differences previously found from examination of the velocity curves are real.

If it is assumed that the secondary is the less massive star, the picture we have of a system such as V Puppis is one consisting of a B1 primary and a slightly smaller B3 secondary, the latter filling its lobe of the first equipotential surface and producing a stream of gas which flows through the inner Lagrangian point towards the following hemisphere of the primary star. There would also appear to be a very short distance between the surface of the primary star and the inner Lagrangian point making it impossible for a gaseous ring to form under these circumstances.

Returning to the case of β Lyrae, despite the vast amount of work which has been done on this particular variable we are still far from arriving at any clear-cut picture of this system which is undoubtedly very different in many respects from the majority of these stars.

β Lyrae is a member of a multiple star system and from a colour-magnitude diagram, an absolute magnitude of −3.9 has been derived for the binary component when at maximum brightness. After making various assumptions and corrections, a mass of about eleven times that of the sun has been found for the primary component, leading to one of more than twenty solar masses for the secondary. Here then it would seem that we have the opposite state of affairs to that prevailing in the Algol variables. However, it must be pointed out that the above value for the mass of the primary does depend upon certain assumptions, particularly that there is no appreciable mass loss from the system (a condition which we have just seen is not the case) and also that the mass-luminosity relationship found for single main sequence stars holds true for the components of a close binary system. We must therefore accept such values with considerable reserve.

One of the visual components of the multiple system of which the binary forms a physical part since they share the same motion through space, is quite definitely a main sequence B6 star and we are probably justified in believing that both it and the binary pair are of approximately the same age. However there are several anomalies in the spectrum of the B-type component of the eclipsing pair and it is logical to assume that these are due to a more rapid evolution of this star due to its proximity to its companion. We are therefore not at liberty to presume that the primary component of the eclipsing system obeys the mass-luminosity law. As an additional confirmation of this it is noted that single stars which have evolved to the right of the main sequence and now lie in about the same position as this star do not exhibit the same spectral anomalies. The figure of eleven solar masses must consequently be viewed with some caution and since the mass of the secondary has been directly derived from that of the primary, this too is suspect.

Possibly a figure much closer to the truth may be found from the assumption (probably correct) that the primary component has filled the inner Lagrangian lobe and the temperature obtained from its spectral class is between 11,800 and 12,800°K.

By this method a somewhat lower value of between 3.5 and 8.5 solar masses is found for the primary, that of the secondary lying between twelve and twenty times the mass of the sun.

These conclusions have, however, not been incontestably established and another and quite opposite view has been put forward. This is that unlike most of the β Lyrae variables, the secondary is the more massive of the two components. There are several very peculiar features about β Lyrae both in its photometric and spectroscopic behaviour which have long puzzled astronomers. For instance, when examined in blue light, the depth of the primary eclipse has been found to vary by 0.3 magnitude in a period of a year or so. This variation in the depth is not observed over a period of a few cycles and taken together these two facts show (a) that it cannot be due to changes in a body the size and mass of a star since these would be expected to take far longer than a year or so and (b) they cannot be caused by changes taking place in a randomly flowing gas stream since variations of this nature would be noticeable from one cycle to the next. However, if we assume the presence of a rotating disc of gas around one of the stars, this could possibly explain the observations.

Secondly the primary eclipse as shown in the light curve is not symmetrical; the decline is markedly steeper than the ascent. An explanation of this is that the gases which are streaming outward from the primary star may produce an obscuration of the system in addition to that produced by the rotating disc. Depending upon the angular momentum, these gases may either disperse away into space or join those already present in the disc.

Another peculiarity of this variable, which has already been discussed, is that of the synchronisation of rotational and orbital velocities. As we have seen, the primary is rotating much too slowly whereas from the extreme closeness of the two stars we would expect them to be synchronised. It has been suggested that because of the rapid evolution of the primary star away from the main sequence, its axial rotation slows as a result of the conservation of angular momentum but from other theoretical considerations it is difficult to reconcile a rotational velocity of only about 40 kilometres per second with the much higher value of about 200 kilometres per second as required for synchronisation, even if we assume that there has been expansion of the primary because of evolution.

In view of this it has been argued that β Lyrae resembles the Algol variables in that the primary and not the secondary is the less massive of the two stars. Such an assumption certainly removes many of the earlier difficulties but poses the question: Why should the secondary star which is more massive be underluminous while the primary, which should have evolved more slowly, has reached the giant stage?

There are three possibilities which have been put forward and which we shall consider here. Firstly it may be that the material from which the secondary star condensed was extremely large thereby retarding the eventual contraction of the star itself. We may therefore postulate a long initial time lag in this star's evolution and indeed, the rotating disc which surrounds this star may be the remains of this cloud of pre-stellar matter.

A second suggestion recently put forward is that because of more rapid evolution, the secondary has evolved through the giant stage and now occupies a position on the Russell-Hertzsprung diagram below the main sequence. This is in accordance with the known fact that the more massive a star is, the more rapidly it will evolve.

Finally, we may envisage such a system as being one in which, to begin with, the primary was the more massive star and consider a probable course of evolution for such a binary system. While the secondary remains on the main sequence, the primary will evolve (being the more massive) becoming a giant, or possibly a supergiant star in the process. One result will be that matter will be ejected through the inner Lagrangian point. This matter will be caught up by the secondary and once such transfer of mass takes place, the distance between the surfaces of the two stars will become smaller causing the equipotential lobe around the primary to shrink thereby increasing the ejection of mass from this star.

Since this process would be an extremely rapid one compared with the lifetime of such a binary system it offers an explanation of the scarcity of variables such as β Lyrae. An important stage in the evolution of such a system will be reached once the masses of the two components are equal since any further ejection of matter from the primary will then make the secondary the more massive of the two giving rise to a situation such as we now find for β Lyrae.

Each of these three possibilities has certain points to commend it and explains fairly plausibly some of the curious anomalies associated with this particular variable. Whatever the final verdict may be regarding these theories, we may readily see that these stars represent an extremely important stage in the evolution of such close binaries.

Spectra of β Lyrae Variables

As will be seen from the data given in Table 2 the β Lyrae variables are, in general, all early type stars. Very few, if any, are known of spectral class F and later. In certain cases, for example S Antliae and UW Canis Majoris, the spectra of both stars may be distinguished while in others, such as β Lyrae and V Crateris, only the spectrum of the primary star is visible.

Mention has been made earlier of the differences found in the orbital elements as derived from the velocity curves. It is now necessary to examine more precisely the reasons for these differences. Since the spectra are of the early types, mainly of class O or B, the lines of both hydrogen and either neutral or singly ionised helium are present and radial velocity curves may be obtained from either of these sets of lines.

In certain of these variables which have been more thoroughly studied it has been found that the radial velocity curves thus obtained are not consistent. The eccentricity and amplitude is not the same in the curves found from the hydrogen and helium lines in the case of the primary star and there is often quite a large difference in the amplitudes of the corresponding curves obtained for the secondary component. In addition, there is a marked difference in eccentricity between the two radial velocity curves found from the helium lines, that for the secondary star being usually less eccentric and more rounded at the peak.

In certain cases, e.g. RY Persei, we know that this difference in amplitude is due to the presence of the gas streams in these systems moving from the primary to the secondary. In others, such as V Puppis, this is not thought to be so and a more likely explanation is that the hydrogen lines of the two components are not completely separated at any phase of the light cycle. Under such conditions it is clearly misleading to try to obtain accurate orbital elements from the radial velocity curves of the hydrogen lines and even if we try to use the absorption lines of helium in the spectrum of the primary star there are complications due to the moving gas stream to take into account. It is now generally accepted that with these particular β Lyrae variables, only the orbital elements obtained from the helium lines in the spectra of the secondary components give values sufficiently consistent to use in the derivation of other parameters.

The radial velocity curve derived from the lines of ionised silicon is shown in Fig. 6. It was from early measurements of the γ-velocity found from this curve that Sahade suggested that there might be a third body in the β Lyrae system which was gradually affecting the γ-velocity but a least-squares solution for the more recent determinations has removed these discrepancies and there is no longer any convincing evidence of changes in the γ-velocities such as would be produced by a third body.

3. W Ursae Majoris Variables

The β Lyrae variables represent one extreme of the class of stars known as the contact binaries, having masses and diameters much larger than those of the sun and periods in excess of half a day. At the other end of the scale we have pairs of stars comparable in size with the sun and forming binary systems so close and compact that their periods of revolution are extremely short, frequently less than 12 hours and sometimes less than 5 hours. Spectroscopically, too, these variables are different; whereas the β Lyrae variables are mostly early-type stars, the class of variable discussed here is generally of type G or later.

Light Curves of W Ursae Majoris Stars

The prototype of this class of star is the eighth magnitude star W Ursae Majoris which was discovered by Müller and Kempf in 1903. The light curves of these variables show minima of almost equal depth and a continuous change of brightness like those of the preceding class and they are generally regarded as a small sub-group of the β Lyrae variables. This latter feature of their light curves indicates that here, too, the components are appreciably distorted into an elliptical shape by powerful tidal and gravitational forces. Indeed, one of the most significant properties of the W Ursae Majoris binary systems is the extremely small separation between the components allowing physical interaction to take place.

Fig. 7 illustrates the mean light curve of W Ursae Majoris.

Detailed analyses of the light curves of these variables have been made by several observers and have revealed that the relative depths of the two minima and also the brightness at maximum can change over a period of time, sometimes even from one cycle to the next. As may be imagined, this makes it extremely difficult to assign any orbital elements to these stars from the light curves alone and much of our present knowledge of these contact binaries comes from a study of their spectra, their combined magnitudes and colour.

Not only are there changes in the shape of the minima but the periods too are found to be variable, that of W Ursae Majoris for example having slowly increased since about 1950.

When examined in both blue and yellow light, further

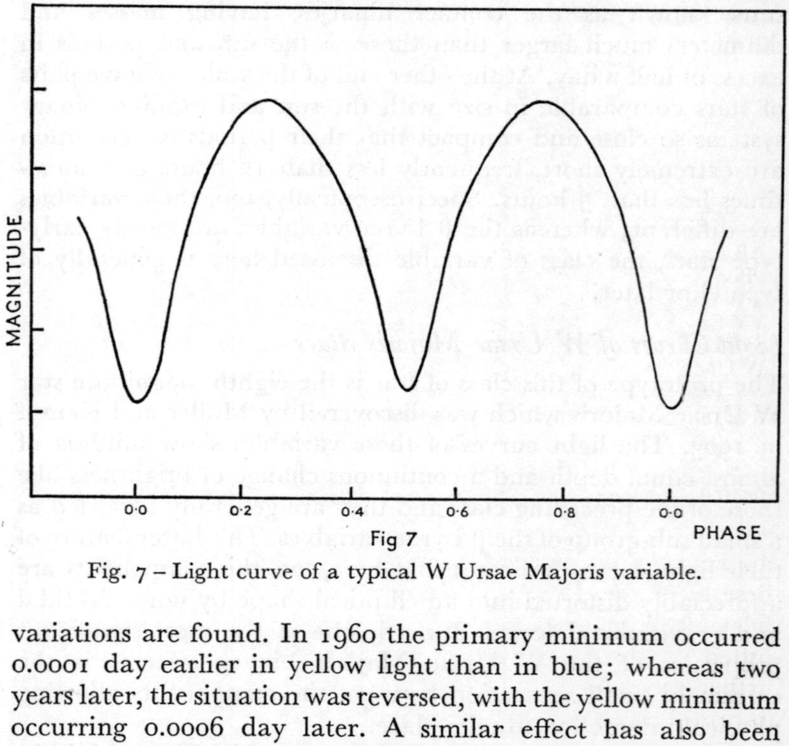

Fig. 7 – Light curve of a typical W Ursae Majoris variable.

variations are found. In 1960 the primary minimum occurred 0.0001 day earlier in yellow light than in blue; whereas two years later, the situation was reversed, with the yellow minimum occurring 0.0006 day later. A similar effect has also been noticed with the times of the secondary minima. Such small deviations are thought to be due to the same phenomenon which causes changes in the shapes of the light curves.

The frequency with which these changes occur has led to the supposition that the photospheres of the individual stars are not of uniform brightness and that there exist areas which are either more or less luminous than the average. Another possibility is that like the β Lyrae variables there exists a belt of matter lying between the two components. This gas may vary in luminosity or may even absorb some of the light of the system. Quite clearly, our knowledge of the conditions existing in these variables is too fragmentary to enable us to determine as yet the true nature of the physical phenomena which are the basic cause of these perturbations in the light of these stars.

Table 3 gives a list of some of the brighter of the known W Ursae Majoris variables.

All of the above magnitudes are photographic.

Table 3

W URSAE MAJORIS VARIABLES

Star	Maximum	Primary Minimum	Secondary Minimum	Period (days)	Spectrum
AB And	9.99	10.93	10.73	0.3318859	G5+G6
i Boo	6.60	7.02	7.02	0.2678080	G2+G2
VW Cep	7.80	8.21	8.14	0.2783199	dG5+dG5
YY Eri	8.37	9.24	9.01	0.3214940	G5+G5
SW Lac	9.25	10.00	9.90	0.3207156	G3p+G3p
W UMa	8.30	9.06	8.98	0.3336370	F9p+F8p

All of the above magnitudes are photographic

Spectra of W Ursae Majoris Variables

Normally, the spectra of the W Ursae Majoris stars show the absorption lines due to both components but the intensities of these lines are often variable, those which are shifted towards the violet end of the spectrum being stronger than the red-shifted ones irrespective of which star is approaching us. A further peculiarity is that whereas the presence of these double absorption lines implies that both stars are of about the same luminosity (and therefore from the mass-luminosity relationship should have about the same mass), the velocity curves generally indicate that one component is about three times more massive than the other. During eclipse, the spectral type too appears to be later and the absorption lines single, regardless of which star is being eclipsed.

Such are, in outline, the general features of both light and spectral changes occurring in this small group of variables. They are much more than an interesting or curious sub-group of the much larger β Lyrae class of variable however; they are important because recently it has been suggested that the W Ursae Majoris stars are the precursors of the U Geminorum variables which will be discussed in a later chapter. If this is so, it might explain the comparative rarity of these objects in the galaxy since if, as the observations tend to show, one of the components is more massive than the other, we would expect it to begin evolving away from the main sequence first becoming hotter and somewhat larger in the process. It is interesting to note in this respect that of the U Geminorum stars which have been more thoroughly investigated, the secondary does turn out to be a yellow dwarf star very similar to those which make up a W Ursae Majoris eclipsing system.

We may therefore postulate that the lifetime of a contact binary of this type during which it exhibits the general characteristics of the W Ursae Majoris binaries will be relatively short accounting for their apparent scarcity. We must not however forget that these dwarf yellow stars are intrinsically quite faint objects and their total number in the galaxy may in reality be quite large, only a few of the nearer ones having been discovered.

4. Peculiar Eclipsing Variables

We now come to those eclipsing systems which cannot be satisfactorily fitted into any of the three preceding classes. It must be appreciated that all of these variables possess either the Algol or the β Lyrae characteristics but with their light curves or spectroscopic qualities sufficiently altered by their spatial environment to warrant their inclusion in a separate chapter.

RY Scuti

One of the first of these peculiar variables to be discovered, by Miss Cannon in 1908, was RY Scuti. Early observers originally came to the conclusion that this star was an irregular variable although Lause noted that the light variations were both large and rapid, the rise and fall sometimes taking only a few days.

In 1922, Merrill undertook a spectroscopic study of this star, showing the presence of five bright nebular lines associated with emission lines of hydrogen and helium previously discovered independently in some nebulae. The radial velocity derived from these bright lines is +30.2 kilometres per second. Attracted by this spectral peculiarity, other observers, in particular Gaposchkin, carried out intensive investigations of RY Scuti proving it to be, not an irregular variable as first thought, but an eclipsing variable with a period of about eleven days. Such a long period is most unusual for such a system particularly one with, as was later shown, elliptical components.

The light curve, as shown in Fig. 8, strongly resembles that of β Lyrae and the periods of the two stars are also similar.

The secondary minimum is well defined in this instance and from the light curve the elements of the system have been obtained. The diameters of the two stars are almost identical being about nineteen times that of the sun and the peculiar spectrum strongly suggests the presence of an associated nebula which is responsible for the bright emission lines although the light variation probably comes wholly from an eclipsing system involved in this nebulosity. The star is situated very close to a recognised dark nebulosity and is not far from another peculiar eclipsing star, W Serpentis.

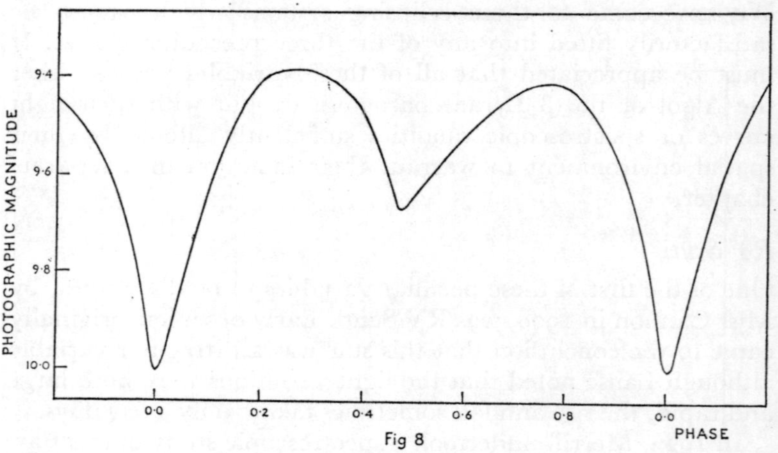

Fig. 8 – Light curve of RY Scuti.

W Serpentis

This particular variable has been the subject of many conflict-ing and puzzling studies since its discovery and it is obvious at once that the light variations are of an extremely unusual type. The mean light curve is shown in Fig. 9 and represents very well the characteristic behaviour of this star.

The generally accepted period, due to Miss Cannon, is 14.15326 days although this was doubled by Zessewitsch but without justification since this merely gives a light curve possessing two identical halves. There appears to be an over-lying longer period which has been determined, with a good deal of uncertainty, as about 270 days but it is unlikely that this period can be regarded as being truly periodic. In addition to the deep minimum there are also two shallow minima and accordingly three maxima. It is noteworthy that the deep minima seem to partake of the long period fluctuation.

Let us now look more closely at the light curve of this particular variable. As will be seen from the mean light curve, the variation appears to be compounded of two kinds, one of the normal eclipsing type and the other as yet undetermined. In the light of our present knowledge of eclipsing variables we may picture W Serpentis as a system in which the two components are of approximately the same size but differ in surface brightness and therefore in total luminosity. The temperature of the fainter star, derived from the depth of primary eclipse, is thought to be around 3,300°K corresponding to a spectral class gK2 and an absolute visual magnitude of −1.5. The respective densities of these stars are then 0.001 and 0.0005 that of the sun, very like a good vacuum here on Earth.

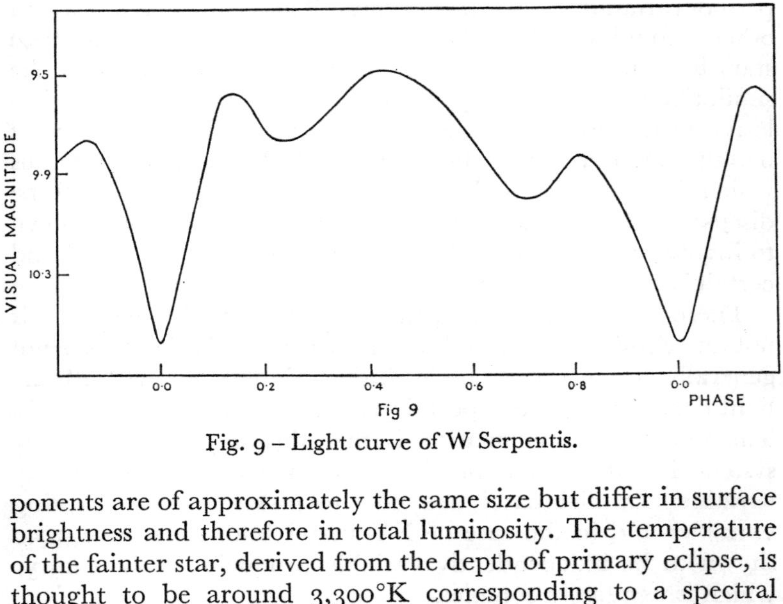

Fig. 9 – Light curve of W Serpentis.

The bright lines of ionised iron in the spectrum of this variable are regarded as originating in a very extensive chromosphere around the brighter star if we assume that this chromosphere is not completely hidden by the fainter companion at the time of a deep minimum. On the other hand there may conceivably be a very extensive envelope surrounding both stars, very similar in nature to the gas streams in the β Lyrae system. As mentioned earlier, W Serpentis is very near to known dark nebulosity and may, in all probability, be at the same distance. If this is so then it will be embedded in this cloud of dark matter whose presence will naturally introduce further parameters which cannot at present be evaluated.

The light fluctuation superimposed on the mean eclipsing curve, producing the secondary maxima and minima, is possibly that of a G-type giant, this being the brighter component of the system. We have also to take into consideration the long period variation of about 270 days. This feature is known in other variables such as DF Cygni and AI Scorpii but these stars have not been recognised as eclipsing variables and the similarity may be illusory. There is, of course, the dark nebulosity near enough to the star to suggest the possibility of nebular interference as in the case of R Coronae Austrinus which is quite definitely an intrinsic variable and will be discussed later in this book but this star has always been shown to involve changes in brightness which are far more rapid and certainly more irregular.

There remains, naturally, the possibility that W Serpentis is not an eclipsing star at all but an intrinsic variable. This is not generally accepted since no intrinsic variable is known with the light curve and spectral peculiarities as shown by W Serpentis. The most likely explanation is that this star is an eclipsing system in which one of the components is independently variable. Parallel cases are not unknown among eclipsing variables and in the case of XY Persei both stars have been shown to be independently variable producing a light curve which is extremely complex.

UX Ursae Majoris Stars

These variables form a small group of stars having Algol-like characteristics but with extremely short periods and certain well-defined differences in their light curves and spectra sufficient to put them into a separate class. One of the stars sometimes included in this group is Nova DQ Herculis (1934) which will be fully described later. The two others which have been thoroughly investigated are RW Trianguli and UX Ursae

Majoris. Both stars may perhaps be old novae but there is no conclusive observational evidence available at the moment to prove this.

Unlike the normal Algol variables, the light curves of these stars show rapid variation of brightness outside of eclipse up to as much as one and a half magnitudes in the case of RW Trianguli, fluctuations which are apparently completely random. These were first noticed in 1958 by Protitch who also discovered a further peculiarity, namely that no evidence of a secondary minimum could be found. During eclipse, the rising branch of the light curve is decidedly asymmetric and at times bright shoulders are visible both before and after minimum.

RW Trianguli also shows one odd feature not present in either UX Ursae Majoris or DQ Herculis namely that the brighter the light curve outside of eclipse, the deeper is the minimum. All of the photometric observations made on RW Trianguli show that the eclipse is quite definitely an occultation unlike that of DQ Herculis.

In all three stars the evidence points to the existence of a stream of gas in the system, part of which is luminous and covered by the dark star during the latter stages of the eclipse, thereby explaining the asymmetry of the rising branch of the light curve towards the end of minimum.

The information available at the present time is in favour of a system in which there is erratic ejection of gas from the small, bright star, this matter increasing the luminosity of the system and surrounding the star itself, effectively increasing its diameter two or three times. It is possible that there may be ejection of gas into the system from the red star but this cannot be proved conclusively although it is known from a study of other close binary stars that the dark companion fills its lobe of the inner Lagrangian surface and loses mass to the hotter star.

The spectrum of RW Trianguli consists of a continuum with strong, narrow lines of hydrogen, helium I and helium II, the latter rather faint. The strengths of the lines are stronger than in UX Ursae Majoris but weaker than in SS Cygni at minimum. The last star belongs to the U Geminorum class of variables which are described in Chapter 16 and which are also close binaries, considered by some astronomers to be closely linked with the W Ursae Majoris stars discussed in the last chapter.

VV Puppis

One of the strangest and most interesting of eclipsing variables is the star VV Puppis. Discovered in 1928 by Van Gent and shown by him to have a period of only 100 minutes, subsequent investigations showed that its light variations are of an extreme-

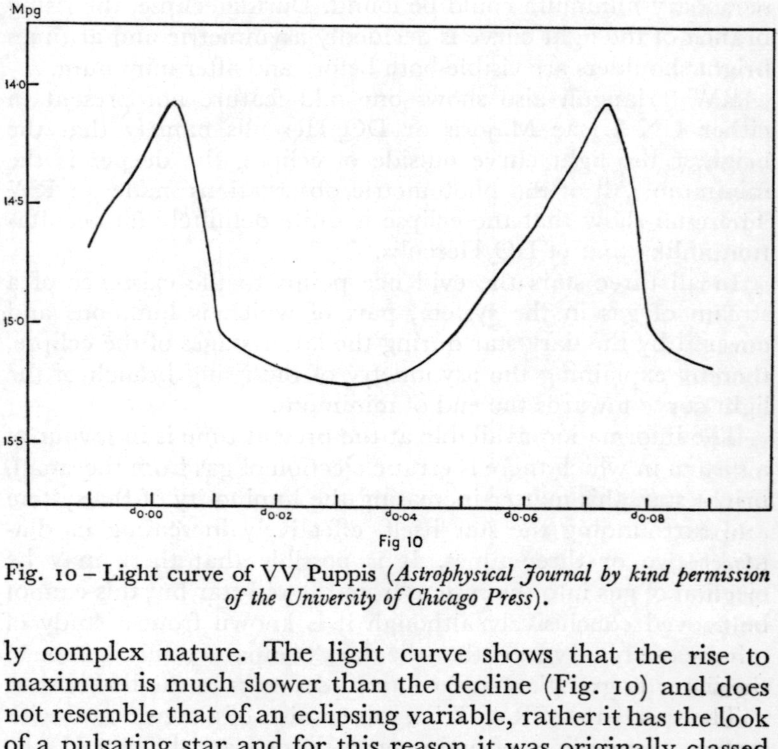

Fig. 10 – Light curve of VV Puppis (*Astrophysical Journal by kind permission of the University of Chicago Press*).

ly complex nature. The light curve shows that the rise to maximum is much slower than the decline (Fig. 10) and does not resemble that of an eclipsing variable, rather it has the look of a pulsating star and for this reason it was originally classed as an RR Lyrae variable of unusually short period.

Since its discovery the star has kept essentially to its mean period of 100 minutes but there have been marked fluctuations, due more to a change in the shape of the maxima than to any real change in the period itself. Superimposed upon the short primary period is a slow secular change. The amplitude of the individual maxima is only about 1.5 magnitudes and at the time of its discovery it varied from 12.8 to 14.4 magnitude, but the mean minimum brightness then faded slowly until in 1949 it became as faint as 17.0 magnitude. It then slowly brightened, the oscillations still amounting to between 1.5 and 2.0 magnitudes until at the present time the mean minimum magnitude is about one to two magnitudes fainter than at discovery. So far there would appear to be no correlation between the two periods.

During the time when the star was very faint it was found from observations made at Pretoria that it exhibited no activity whatever although it is quite possible that it faded at intervals to less than 17.0 magnitude at this time. Then, after about a month, the periodic fluctuations were found to begin again, faint at first, but building up slowly to the normal amplitude. Such abnormal behaviour is quite unknown among the RR Lyrae variables and a more likely explanation is that this star is, in reality, a binary and not a pulsating object.

The spectrum of this remarkable star shows strong emission lines of hydrogen and helium II and is, in fact, very similar to those of Nova DQ Herculis (1934) and to RW Trianguli and UX Ursae Majoris described in the previous section. A radial velocity curve has been obtained from Hβ, Hγ, Hδ and He II, indicating a velocity variation of approximately 900 kilometres per second. On the assumption that VV Puppis is an eclipsing system, it has been suggested that unlike UX Ursae Majoris and similar variables the larger component is the brighter of the two and the surface luminosity of this star is not uniform causing the variations in brightness from one cycle to the next. This non-uniformity of brightness in the large star is regarded as changing slowly with time producing the slow secular changes which are superimposed on the shorter cycle and at those periods when the star is extremely faint and there appears to be no variability of light, the bright spot on the surface of the primary has faded almost completely.

We thus have here an almost unique system although, as has been pointed out by Herbig, there are other alternatives to this picture of VV Puppis and no final decision can be reached without much further intensive study of this most remarkable star.

Binary Systems Among Old Novae

Although the novae will be fully described in a later chapter, it will be as well here to examine the physical characteristics of some old novae at minimum light where there is evidence for eclipse, providing us with important evidence that they are binary systems. The first of these old novae to be shown to be a binary was Nova DQ Herculis which reached a maximum of 1.3 magnitude in 1934, subsequently fading to around fifteenth magnitude about three years later. In 1954 Walker showed that in its period, physical properties and spectral characteristics it is not unlike the U Geminorum variables described in Chapter 16 where it will be seen that most, if not all, of these are double star systems.

The periods of these old novae are extremely short. That of DQ Herculis is only 4 hours and 39 minutes and clearly the two components are very close together. In the spectrum, the lines of helium II and also the higher lines of the Balmer series of hydrogen show a variation in orbital velocity which is not shared by the lower members of the Balmer series or those of silicon II and oxygen II. This observation lends weight to the suggestion that there are two distinct regions in the system which produce the emission lines; one a high density disc or shell which moves along with the blue primary in its orbit and the other a lower density shell which is expending slowly. The spectrum of the secondary companion is not visible either during or outside of the eclipse.

Greenstein and Kraft have also demonstrated that the radial velocity curve of helium II shows a rotational disturbance beginning and ending with the eclipse. One must therefore conclude that here we are dealing, not with a normal eclipse of the stars, but with a semi-transparent ring surrounding the hot component. This star has an extremely high surface temperature and much of its radiation lies in the ultra-violet region of the spectrum. It is this ultra-violet radiation which causes the gases in the enveloping ring to fluoresce giving rise to a hydrogen emission which is responsible for the blue and visible light of the system. Calculations show that the primary is a bright, white dwarf having a mass only 0.12 times that of the sun, the secondary being a red star of only 0.20 solar mass.

Several other old novae have been discovered to be binaries, including GK Persei (1901), T Aurigae (1891), V603 Aquilae (1918) and also the recurrent novae T Coronae Borealis (1866 and 1946) and WZ Sagittae (1913 and 1946). Of other such stars which have been investigated, the spectrum of the recurrent nova V1017 Sagittarii (1901 and 1919) has been found to be composite, the spectrum being that of a blue continuum superimposed upon a fairly well-developed absorption spectrum of a late-type star, possibly one of class G5p.

A further recurrent nova which has been extensively studied is the well-known variable RS Ophiuchi. At minimum light an absorption spectrum is present in addition to several bright emission lines but from various considerations it would appear that the former originate in a shall of gas rather than in a stellar object.

Emerging from the above discussion we can build up a picture of an old nova consisting of a small blue or white star which is the seat of the original nova outburst and a much larger red companion and gradually, as more information becomes available, especially spectra obtained with large telescopes (for the majority of these old novae are extremely faint objects), many astronomers are coming round to the idea that one necessary condition for a star to experience a nova outburst is that it must initially be a member of a close binary system of a very particular kind, namely a blue and a red star with a stream of gas flowing from the red star to the blue one and forming a shell or disc around the latter giving rise to the observed bright emission lines which from observation indicate that this gaseous shell moves with the blue star in its orbit. It is even conceivable that this stream of high temperature hydrogen gas may be the ultimate cause of the nova explosion by reacting, in some as yet unknown manner, with the helium core of the blue companion.

At this point it may be mentioned that certain of the old nova possess some of the distinguishing characteristics of the U

Geminorum variables in that minor outbursts of light have been observed over the years since the original explosion. GK Persei for instance has brightened by one or two magnitudes on several occasions. Twice during 1967 it has attained eleventh magnitude, these maxima being very reminiscent of the behaviour of SS Cygni and other U Geminorum variables. Quite obviously, the novae are far from quiescent objects once they have faded to minimum.

Wolf-Rayet Binaries

In the Introduction mention was made of the stars of spectral type O which form the beginning of the Harvard sequence of stellar spectra. These are all extremely hot stars, their spectra characterised by the presence of lines due to ionised helium and those of oxygen and nitrogen which have been doubly and triply ionised, a condition indicative of the very high surface temperature which must be between 30,000 and 40,000°K. Almost a century ago, two French astronomers, Wolf and Rayet, discovered certain stars whose spectra, although similar to the normal type O stars, contained several wide emission bands. The surface temperature of these Wolf-Rayet stars are far in excess of those normally classed as O stars, possibly as high as 80,000 or 100,000°K, and a special type W has now been created to accommodate them.

Although they are so hot and therefore highly luminous objects, very few of these stars are known. From an analysis of their spectra it is possible however to distinguish two sub-classes designated as WN and WC respectively, the former having spectra dominated by emission lines of ionised nitrogen and the latter by bands due to carbon. We may look upon this sub-division as being somewhat similar to that of the class N and R stars although at the other extreme of the temperature scale.

Owing to their rarity, we would not expect to find many eclipsing variables in which one star is a member of this class and indeed only four eclipsing binaries are known in which one component is a Wolf-Rayet star. These four variables are CQ Cephei, CX Cephei, V444 Cygni and CV Serpentis. A fifth star known as HD 211853 is a Wolf-Rayet spectroscopic binary having a period of 6.6864 days but as such binaries are not true variable stars it will not be discussed further. Of the four, only CV Serpentis contains a WC star in the system, the other Wolf-Rayet stars all having spectra of the nitrogen sequence.

The light curve of CQ Cephei is of the W Ursae Majoris type with the difference that the eclipses are asymmetrical and there are irregularities which occur from cycle to cycle. This tendency for a lack of consistency between individual cycles is probably a common feature in the light curves of these four stars and is almost certainly due to intrinsic variations within the systems themselves.

CV Serpentis has been investigated by several observers with somewhat discordant results. Gaposchkin found that the primary and secondary minima are only 0.14 and 0.08 magnitude deep respectively requiring sensitive photoelectric equipment for their measurement. In a more recent investigation, Hjellming and Hiltner found that the depth of the primary minimum is much greater than that previously discovered, being 0.55 magnitude in blue light and 0.49 magnitude in the yellow. The primary minimum found in both investigations was shown to be decidedly asymmetrical. Assuming that the orbit is circular, the derived period for this variable is 29.640 days.

Elliptical Variables

So far in this discussion we have been dealing with binary systems in which an actual eclipse occurs, either total, annular or partial depending upon the diameter of the components and the inclination of their orbit to our line of sight. It has been tacitly assumed that where the systems have a high inclination of the orbit no variation in light will occur since there will then be no actual eclipse. Generally speaking this is a valid assumption but where the stars are pronouncedly distorted by tidal and gravitational forces there is the possibility of light variation due to continuous changes in the surface area as they revolve around their common centre of gravity. Such changes in brightness will inevitably be extremely small.

Nevertheless, five such variables are known at the present

time, three of them being in the constellation of Perseus, the other two being ζ Andromedae and π^5 Orionis. Their amplitudes are between 0.06 and 0.1 magnitude and accordingly their light changes can only be followed photoelectrically.

PART TWO

Intrinsic Variables

5. RR Lyrae Variables

Unlike the eclipsing variables discussed in Part One, the intrinsic variables are, for the most part, single stars and not multiple systems, owing the variations in their brightness to changes which actually take place within the star itself, generally speaking in the outermost layers. The division of the intrinsic variables into pulsating and eruptive stars has been obvious and accepted by astronomers for many years but as we shall see the boundaries between certain of the classes are far from clear and many groups contain very dissimilar stars so that further reclassification will inevitably become necessary as our knowledge of these variables grows.

One of the largest classes among the pulsating variables is the Cepheids named after δ Cephei, the first of this class to be discovered and also one of the brightest. The light cycle of the vast majority of these stars is repeated faithfully from one maximum to the next, the period remaining constant, often to within a fraction of a second.

The Cepheids form an extremely important group of variables since, as will be seen later, they provide astronomers with one of the few direct methods of measuring not only interstellar, but also intergalactic, distances with a high degree of precision. In practice it has been found that the periods of the Cepheid variables fall into two well-defined groups (Fig. 11); those having periods of less than about one day and those whose periods are in excess of 1.5 days. It seems to have been firmly established that there are virtually none of these pulsating stars with periods of between one and one and a half days and for this reason the class is divided into two main groups – the RR Lyrae or Cluster variables and the classical Cepheids depending upon their periods. We shall later see that this is more than just a matter of convenience since there are now strong reasons for believing that these two classes of variable are physically different.

In the present chapter we shall be discussing the RR Lyrae stars which form an interesting group in themselves. Until the early part of this century only a few of these variables were known in our galaxy although large numbers had been found in the globular clusters which are huge, compact groups of stars forming a kind of halo around the galaxy. It is from this

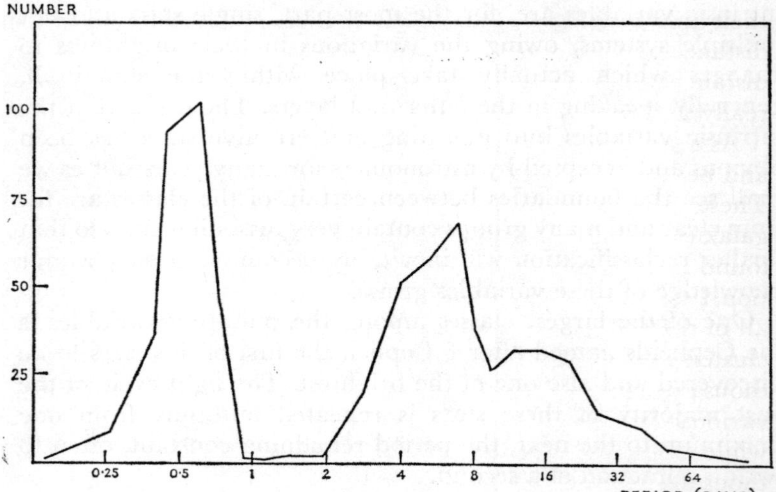

Fig. 11 – The period-distribution curve for RR Lyrae and Cepheid variables.

that their name of Cluster variables derives. Since that time, however, more than two thousand have been discovered within the confines of the galaxy and are being used by astronomers for the solution of a wide variety of stellar astronomical problems.

To the layman the measurement of interstellar distances may appear a completely impossible task. When one considers that the distance to even the nearest star is so great that it takes light, travelling at more than 186,300 miles every second, four and a third years to reach us, the problem of measuring the dimensions of the galaxy seems utterly out of the question and becomes even more incapable of solution when we attempt to estimate the distances of the extragalactic nebulae which are found to be several millions of light years away.

Nevertheless, the methods used by astronomers are really quite simple in practice. One of these, using the RR Lyrae stars as indicators, depends upon a well-established fact which has been known for several centuries, namely that the brightness of a light source is inversely proportional to the square of its distance. Now it has been found that all of the RR Lyrae variables have almost the same intrinsic luminosity and from their apparent brightness it is therefore possible to estimate their distance. Shapley has utilised this property to determine the distances of many of the globular clusters providing the first really satisfactory model of the shape and size of the galaxy.

As a general rule these variables possess large radial velocities and belong to what is known as the Population II class of stars. These are those found in the globular clusters, elliptical galaxies and the nuclei of the galaxies. By contrast, the stars found in the spiral arms of the galaxies are termed the Population I stars. The main point of difference between them is that all of the Population II stars are old. The ages of the globular clusters for example are estimated at between five and six thousand million years whereas those of Population I are of various ages, some not much more than a few million years old while others, such as the sun itself, came into being about four thousand million years ago.

One point of interest is that very many faint RR Lyrae variables have been discovered at great distances; evidently they are either at the outermost periphery of the galaxy or perhaps even in intergalactic space.

The Classification of RR Lyrae Variables

The first really serious attempt to classify these stars was made by Bailey who showed that they could be conveniently divided into three groups according to their period and the general characteristics of the light curve.

Stars of Group **a** have relatively large amplitudes and periods of the order of 0.5 day. The light curves show a rapid rise to maximum (which is often quite sharp) followed by a slower decline with the minimum almost flat and undisturbed and lasting for almost half of the total period. RR Lyrae itself belongs to this particular group.

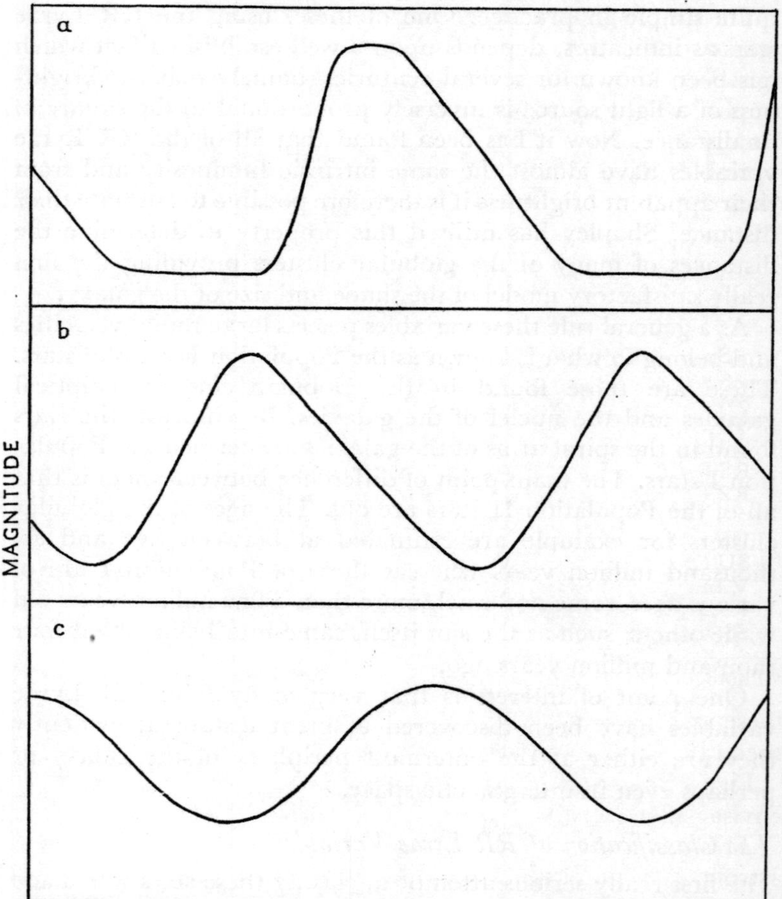

Fig. 12 – Typical light curves of the Bailey-type RR Lyrae variables.

Group **b** stars are very similar but the amplitudes are found to be smaller and there is a somewhat slower rise to maximum which quite often is almost flat. These are variables which remain at minimum for only a small fraction of the period. Like those of Group **a**, however, their light curves are still asymmetrical.

The Group **c** variables on the other hand have quite different light curves which are almost symmetrical with a continuous light variation. Their periods are about 0.3 day and their amplitudes are smaller than in the two preceding classes being usually less than half a magnitude. RV Coronae Borealis is typical of this group of variables. Because of the sinosoidal nature of their light curves some confusion can arise between

Table 4
RR LYRAE VARIABLES

Star	Maximum	Minimum	Period (days)	Spectrum
SW And	9.30	10.31	0.44227818	A3 – F8
SW Aqr	10.64*	12.42	0.45930234	A3
S Ara	10.20*	11.68	0.45188817	A3 – F5
RS Boo	9.71*	11.22	0.37733657	B8 – F0
RZ Cep	9.20*	9.80	0.30867872	A0 – A3
UY Cyg	10.40	11.40	0.56070478	A5 – F0
SU Dra	9.40*	10.40	0.66041926	A2 – A5
V Ind	9.22*	10.22	0.47960255	A0 – F2
RR Lyr	7.13*	8.03	0.56683500	A2 – F0
SS Tau	11.80	12.70	0.36996870	A0 – F3

* Photographic magnitude

these stars and the W Ursae Majoris eclipsing variables if the spectral type and radial velocity is not known.

Fig. 12 illustrates the three groups of RR Lyrae stars as envisaged by Bailey.

This particular classification is still used extensively by many astronomers although another scheme has recently been devised which takes into account the effect of the Blazko Phenomenon mentioned below and which will be discussed in detail later in the chapter.

During the early investigations of the RR Lyrae stars towards the end of the last century it was assumed that, like the great majority of the classical Cepheids, their periods and amplitudes remained essentially constant from cycle to cycle.

It must be borne in mind however that at this time photoelectric techniques had not been sufficiently developed to allow measurement of small variations in the light changes to be made with any accuracy and it was not until the advent of the more refined methods of magnitude measurement and detailed studies of these stars became available, that it was recognised that only a small proportion of them do have constant periods and amplitudes.

Between 1905 and 1908 Blazko found from a long and detailed study of the variables RW Draconis and XZ Cygni that both the shape and amplitude of their light curves were varying with periods of 41.72 and 57.24 days respectively. The Blazko Phenomenon has since been found in a large number of RR Lyrae stars and as yet the cause is not fully understood. It was originally suggested that this variation may be due to the interference of two or more oscillations taking place within the mass of the star but more recent work has indicated that the true picture is even more complex than this.

Like the classical Cepheids which they resemble closely in many respects, the RR Lyrae variables change not only in brightness but also in spectral type, radial velocity, temperature, size and colour index throughout their light cycles. Generally speaking, maximum brightness corresponds with maximum velocity of approach as measured by the Doppler shift in the spectrum lines and also with maximum temperature as a result of the star's pulsation, but there are also shock wave phenomena within the star itself which must be taken into account. The variation in the size of the star which we must postulate to account for the changes in the brightness is not large, only about a tenth of the whole diameter. Indeed, it is doubtful if the entire star swells and contracts during these pulsations. Far more likely it is only the outer layers of the atmosphere which are involved in the volume changes although we must await further work on the theory of pulsation in stars before this is finally confirmed.

One very important fact has recently emerged from a study of the spectra of these variables; there is a distinct difference between the spectral class as deduced from the hydrogen lines and that found from the calcium lines in the same spectrum.

All stars which belong to Bailey's Groups **a** and **b** are approximately of class F5 at minimum as judged from the absorption lines of hydrogen whereas from an examination of the strength of the calcium lines the spectral class is often earlier. It is now generally accepted that the hydrogen line spectral type gives an indication of the surface temperature of the star while the *difference* between it and the calcium line spectral type is an expression of the abundance of metals present in the atmosphere of the star.

Following a systematic examination of the spectra of more than 200 RR Lyrae stars, particularly by Zessewitch, it was shown that those stars which have a variable period are poorer in metals than those which possess constant periods and the former have, on the average, longer periods than the latter. From this it would seem that the stability of the pulsations of these variables is dependent in some, as yet unknown, way upon the relative abundance of metals in the outer layers of the star's atmosphere.

At this point, taking into consideration the factors mentioned above, it is convenient to divide the RR Lyrae stars into four classes:

Class A: Short period stars which show the Blazko Phenomenon quite pronouncedly but which have superimposed secondary periods only a few times longer than the primary fluctuations so that the overall periods do not exhibit marked progressive changes with time. Stars which are typical of this particular class are DY Pegasi, SX Phoenicis and CY Aquarii.

Class B: These are variables which correspond almost exclusively to Bailey's group **c** having almost sinusoidal light curves, quite small amplitudes and periods lying between 0.1 and 0.45 day. Stars which have variable periods such as RU Piscium and T Sextantis, and others with constant periods, e.g. CW Serpentis and AU Virginis, are both placed in this class.

Class C: Stars with periods between 0.25 and 0.75 day. As in the previous class, variables with constant or varying periods are known and many, although not all, exhibit the Blazko Phenomenon.

Class D: These are the long period RR Lyrae variables with

periods in excess of one day. The light curves show large amplitudes and are generally highly asymmetrical. Typical stars of this class are VZ Aquilae, CE Herculis and BF Serpentis.

RR Lyrae

Since the beginning of the century many astronomers have made long and detailed studies of the magnitude changes of this particular variable visually, photographically and photoelectrically. The fundamental period is one of 0.566835 day but on top of this is a much longer period of 40.81 days or about seventy-two of the primary cycles with the general form of the light curve changing progressively throughout this period. In 1926, Prager demonstrated the existence of yet a third period, a very slow variation, confirmed by Miss Fringant as one of approximately 9.6 years. There may even be a fourth period for this star since Walraven has suggested from photoelectric observations that there is a median period equal to about three of the 40.81 day cycles.

By means of a special spectroscopic technique devised by Chalonge, Miss Fringant has been able to show that RR Lyrae has an absolute visual magnitude of +0.8 thereby confirming other evidence that these variables are not nearly as luminous as was originally supposed when an absolute visual magnitude of 0.0 was assigned to them. As mentioned earlier these stars have been extensively used to measure interstellar distances and in particular those of the globular clusters and from them the size of the galaxy. This more recently determined value has had the effect of reducing the distances thus deduced for the dimensions of the galaxy.

We do not yet fully understand the basic cause of the oscillation in brightness of these stars but since the pulsation theory was first proposed by Shapley and Eddington, a tremendous amount of theoretical work has been carried out and it now appears that pulsation is connected quite critically with the structure of the star. To understand this we must first consider how a star evolves and the different evolutionary tracks it may follow depending upon its mass.

The Russell-Hertzsprung Diagram

There are two major characteristics of any star which are fairly readily observable. These are:

(a) The brightness of the star – for variables we may define this as the amount of visible light which is emitted at maximum.

(b) The surface temperature which can easily be determined from the distribution of light with respect to wavelength in the spectrum.

When the brightness is plotted against surface temperature for a large number of stars which are essentially at the same distance from us such as those in a particular globular cluster, we obtain the now familiar Russell-Hertzsprung diagram which is shown in Fig. 13, the positions of certain classes of variable star being also shown.

The continuous line is known as the main sequence and all stars which lie on or very close to it are known as the main sequence stars. These are stars which contain different amounts of material in them but the composition of which is the same for them all and is actually the same as that which makes up the sun which is also a main sequence star. Although this is perhaps the simplest case of all it is neither irrelevant nor insignificant since it represents the conditions prevailing at the time when the stars were born out of the interstellar gas and dust clouds. The position of a star on the main sequence is therefore determined by its mass and it will remain on the main sequence so long, and only so long, as certain nuclear reactions are going on inside it, providing it with the necessary energy to keep shining.

The basic nuclear process which provides the initial energy for the star after condensing out of the gas cloud is known as the proton chain reaction. When a star condenses it is composed almost entirely of hydrogen which has a nucleus consisting of a single particle known as a proton. At the low temperature and density prevailing under these conditions, the only possible reaction is that of a proton combining with another proton to form a deuteron, a particle in which the nucleus is made up of a proton and a neutron. In this process, one of the protons changes into a neutron and a positive electron, or positron, is emitted. A further proton is then added to give an isotope of helium, radiation being emitted at this stage; and finally, two

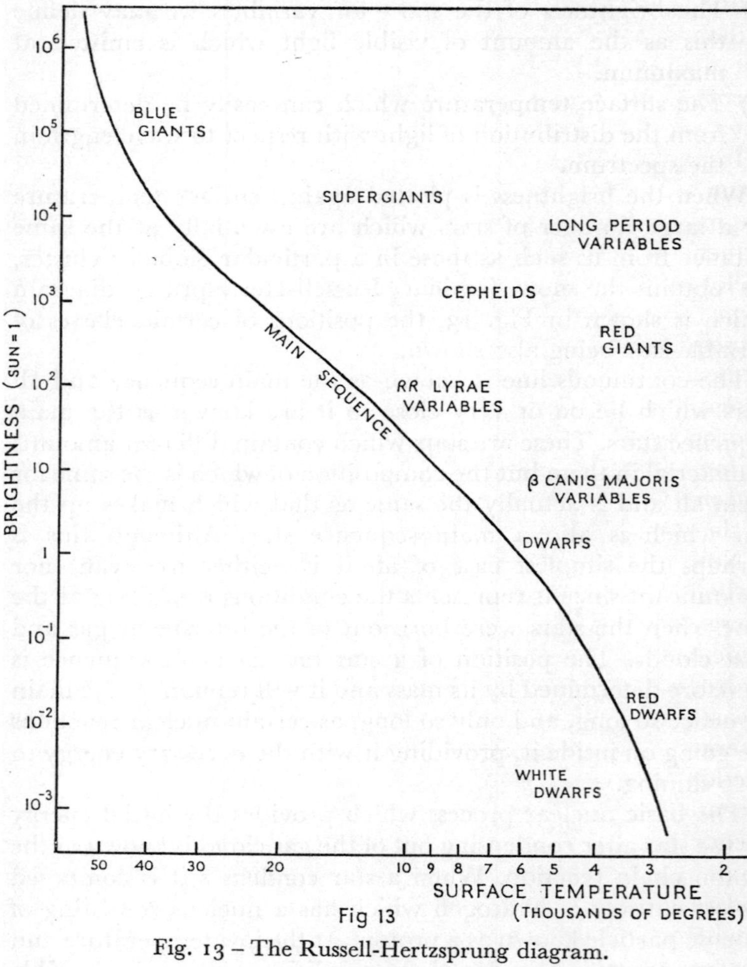

Fig. 13 – The Russell-Hertzsprung diagram.

such nuclei combine to form ordinary helium with the ejection from the nucleus of two protons which are then capable of continuing the cycle all over again. The net result of this series of reactions is the conversion of hydrogen into helium with the production of energy which appears in three forms – the positron, the radiation and the energy of motion of the two ejected protons. Another particle known as a neutrino is also emitted in the first stage of the cycle but as this leaves the star without hindrance or further reaction its energy is completely lost into space. We shall see later in the book that these peculiar particles play an important role in the mechanism by which a nova outburst occurs.

Inside those stars which have a luminosity not more than about ten times that of the sun, the proton chain reactions are the most important. In the brighter stars, however, a different sequence of nuclear reactions known as the carbon-nitrogen cycle attain a far more prominent role. The reactions which form the carbon-nitrogen cycle were discovered by Bethe and subsequently modified slightly by Fowler. In this sequence there are six steps which will not be gone into in detail here. Essentially, the overall reaction is again the conversion of hydrogen into helium with isotopes of carbon and nitrogen acting as nuclear catalysts, i.e. they are both initial reactants and products of the sequence and are used over and over again in the process.

Now regardless of whether the proton chain reaction or the carbon-nitrogen cycle is the predominant means of producing energy, the resulting change in the composition of the star is the same, namely hydrogen is built up into helium. The amount of hydrogen decreases with time while the helium content increases accordingly. Originally it was assumed that in the interior of a star the helium is thoroughly mixed due to convection currents which slowly circulate within it. More recently, however, serious doubts have been cast upon this theory and it now seems highly improbable that there is any mixing at all. Mestel has investigated the problem thoroughly and reached the conclusion that the only point of uncertainty is whether magnetic fields inside a star may allow some small degree of mixing.

Theoretically it can be shown that if there is a complete mixing of hydrogen and helium within a star it will evolve very slowly to the left of the main sequence as time goes on. In the much more probable case where there is no mixing, the evolution of the star will be along a line to the right of the main sequence as shown in Fig. 14.

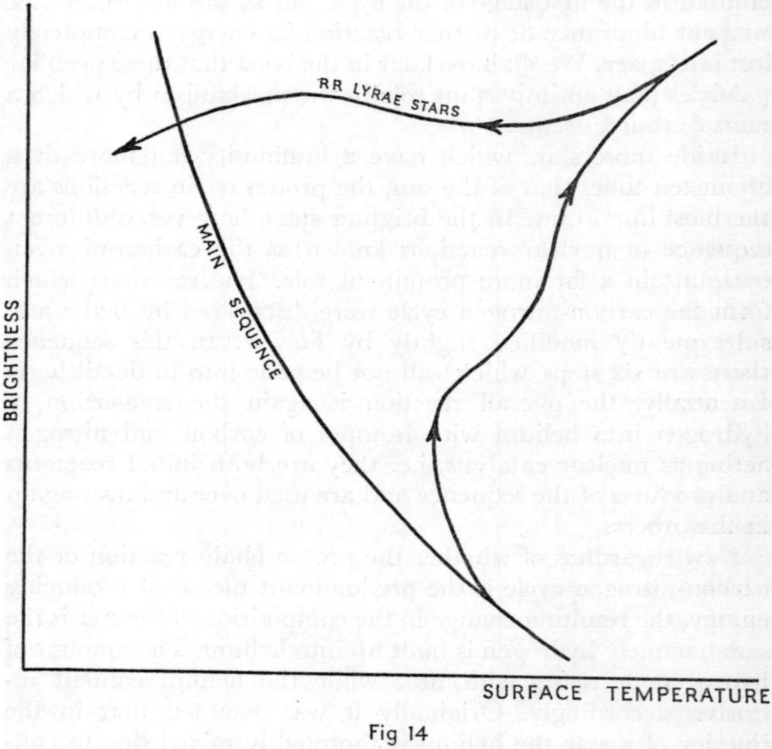

Fig. 14 – The curve of evolution of an unmixed star.

Stars with masses not very different from that of the sun all take similar paths of evolution, becoming brighter and larger until they reach the kink in the curve when a shrinkage takes place, the star gradually getting fainter until it becomes a blue star with a size not much different from that which it had before it began evolving away from the main sequence.

During the early stages of its evolution, helium is produced more rapidly in the centre of the star. This is not unexpected since it is here that both temperature and density are greatest. After a time, however, the star will develop a core of helium and energy will then cease to be produced in the centre. Gamow and Critchfield have shown that when this occurs, a skin is produced around the core which then shrinks to raise the temperature sufficiently for other nuclear reactions to commence inside this skin.

After a time the shrinkage stops otherwise the pressure in the central regions would become so high that a new form of pressure known as degeneracy pressure would develop and become so strong that it would overcome the weight of the overlying layers of gas. A condition of equilibrium must therefore be set up and this has been found to depend quite critically upon an extremely important fact discovered by Chandrasekhar, namely that the mass of the helium core must not exceed 1.44 times the mass of the sun. When we come to discuss the supernovae we shall see what happens if the mass of this helium core should exceed Chandrasekhar's Limit.

In Fig. 14 it will be seen that the RR Lyrae stars occupy a particular point on the evolutionary curve. Although we still do not know for certain the cause of the characteristic oscillations of the variables, we do know that stars on either side of this critical point on the curve do not oscillate while those which are just inside it do. The changeover is very sharp and a significant transition.

A further point of curiosity about the RR Lyrae variables is that although they occur in great numbers in the globular clusters, the proportion of these stars differs widely in different clusters. Between 150 and 200 of them are known for example in Messier 3, but in Messier 92 there are only half a dozen or so known.

The reason for this wide divergence in numbers is not yet known although a statistical survey covering several of these clusters indicates that the former is unusually rich in this type of star.

Proper Motions of RR Lyrae Stars

The proper motions of about two hundred of these variables are known at present but unfortunately most of the early determinations of position were made from photographic plates taken for the compilation of the various astrographic catalogues resulting in a poor quality image and an error which was difficult to compensate for. In addition where the change of position is measured relative to nearby comparison stars on the photograph this will only give a relative proper motion and it is extremely difficult to correct this to provide absolute proper motions since the other stars will also be changing their positions with time. We are therefore forced to accept the fact that a large proportion of the earlier determinations are subject to error.

Several astronomers are now engaged in the task of redetermining these proper motions which are mostly exceedingly small and subject to quite serious systematic errors which will not be eliminated until all of the measured proper motions have been reduced to a consistent system, especially those which have so far only been measured over a short interval of time. Since, in general, they are very distant objects, accuracy demands quite a long interval between determinations of position.

Fortunately there appears to be a wide dispersion in the proper motions of this class of variable and statistical analyses tend to reduce certain of the inherent inaccuracies of measurement. If we can assume that these stars all possess the same absolute visual magnitude then we may derive mean distances from their mean proper motions and also mean radial velocities.

Apparent Magnitudes

The most accurate method for measuring apparent magnitudes is without doubt the photoelectric cell. Visual estimates are all subject to personal error and photographic determinations depend greatly upon the colour of the star; indeed, many of the published photographic magnitudes of these stars show an error of as much as half a magnitude. Photoelectric magnitudes are available for about a hundred variables it is true and were the RR Lyrae stars a strictly homogenous kinematic group this would be quite sufficient data for a statistical analysis to be made, but as we shall see from the radial velocities of these stars, this is not the case. As a result, when we subdivide these variables further into two or more groups there is really not enough data on which to base a satisfactory statistical analysis within each individual group.

One further note of caution must be introduced here when we attempt to obtain distances from the relationship between apparent and absolute magnitudes. Contrary to what might be expected, interstellar space is not empty. The presence of an extremely rarified gas between the stars has been demonstrated by the presence in the spectra of very distant, hot stars of sharp absorption lines due to ionised calcium. Hartmann, at the beginning of the century, observed the K line of ionised calcium in the spectrum of the spectroscopic binary δ Orionis and found that whereas the lines of hydrogen and helium showed the characteristic Doppler shift due to the motion of the component stars around their common centre of gravity, the K line remained stationary proving that the calcium was not present in the atmosphere of the star itself but lay between us and the star. Since that time, the stationary line of ionised calcium has been observed in a very large number of stars. At first it was considered that this very diffuse cloud of gas lay quite close to the sun but we now know that it pervades most of the galaxy. Undoubtedly other elements go to make up the composition of this interstellar gas and according to recent calculations it would appear that the greater part of it is composed of hydrogen but this is extremely difficult to prove spectroscopically.

Most of this gas lies along the plane of the galaxy in the direction of the Milky Way (accounting for the fact that the number of extragalactic nebulae appears to decrease rapidly as we move towards lower galactic latitudes) and it has the inevitable effect of dimming the light of the stars as we go progressively from the poles towards the galactic plane. A further difficulty is that this interstellar gas absorbs selectively in the blue and violet so that there is a progressive reddening effect which has also to be taken into account when estimating the apparent magnitudes of the more distant stars. It is not easy to determine the total absorption produced by this gas but the problem is not an impossible one. From a determination of the mean colours of the RR Lyrae stars which are found at high galactic latitudes and which are therefore only slightly reddened, we may estimate the magnitude effect with a fair degree of precision and use this result to find the absorption in the case of stars at low galactic latitudes.

Light Curves of RR Lyrae Variables

Following Bailey's pioneering work it is now firmly established that the RR Lyrae variables do not form a homogenous group as far as their light curves are concerned. Not only are there the three Bailey classes, **a**, **b** and **c**, but these stars may be further subdivided into those which are singly periodic and those, such

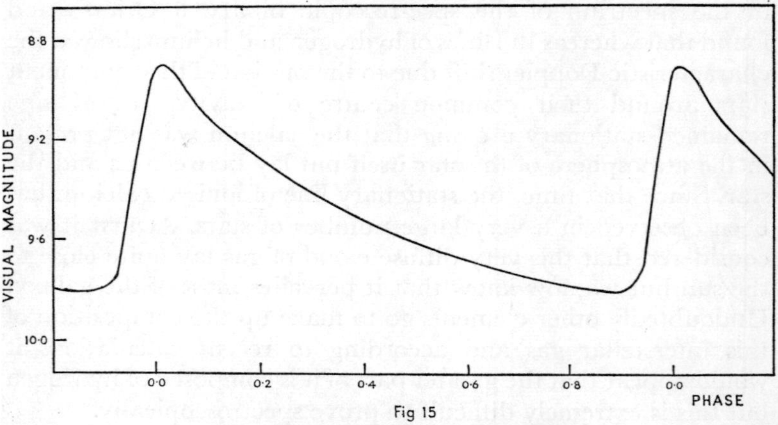

Fig. 15 – Light curve of X Arietis.

as RR Lyrae itself, which has at least three fairly well-defined periods.

The steep rise to maximum which is typical of Bailey's class **a** variables is clearly shown in the light curve of X Arietis. (Fig. 15.)

The visual light curve as illustrated above does not define the slight hump on the ascending branch which is visible only on curves drawn from the far more accurate observations obtained photoelectrically. The large majority of class **a** variables show such a hump. One star however in which it is conspicuously absent is SW Aquarii. Those variables which have a long period superimposed upon the primary one are variable in this respect, exhibiting humps during certain phases of the secondary cycle and not at others.

The presence of a more or less conspicuous hump on the ascent to maximum may therefore be regarded as quite a common feature of the class **a** RR Lyrae stars although by no means an invariable one.

The class **b** variables (Fig. 16) with their much slower rise to maximum form a group intermediate between types **a** and **c**. Very often it is difficult simply from an examination of the

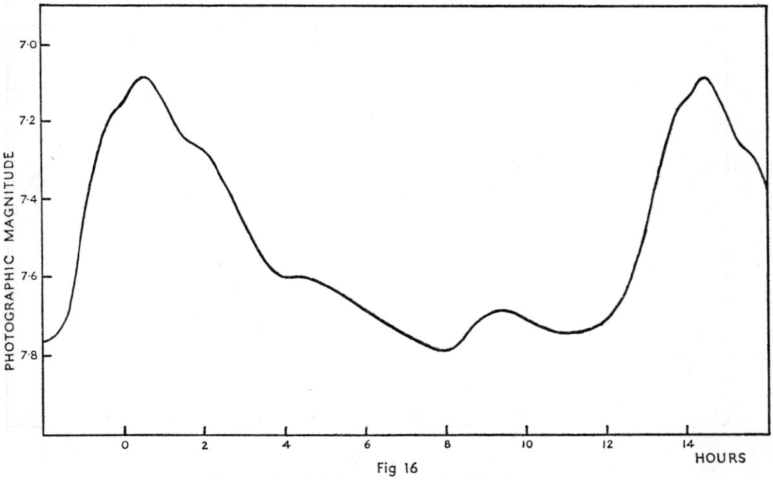

Fig. 16 – Light curve of RR Lyrae.

light curve to place a star with any degree of certainty in this class.

On the other hand, the light curves of the Bailey type **c** stars are quite distinctive (Fig. 17), and like the type **a** variables often show a hump on the ascending branch, frequently just before reaching maximum brightness. The sinusoidal nature of their light curves is well illustrated in that of T Sextantis.

Spectra of RR Lyrae Variables

In the Bailey type **a** variables, hydrogen lines appear both in emission and absorption but not at all phases of the light variation. The emission lines of hydrogen are transient features in the spectra of these stars occurring about midway from minimum to maximum light. No emission lines of hydrogen have been observed in the spectra of classes **b** and **c** variables. One particularly remarkable feature of these emission lines is that in those stars where they are abnormally weak only one component, displaced towards the violet, is seen whereas in variables showing strong emission, two components on either side of the absorption line are visible. The bright emission lines are not of long duration, lasting for only a matter of ten to

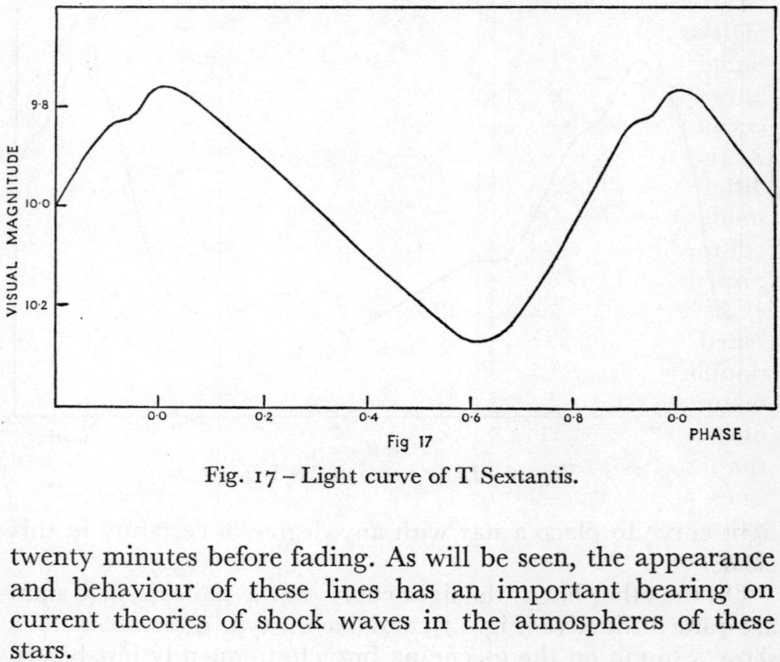

Fig. 17 – Light curve of T Sextantis.

twenty minutes before fading. As will be seen, the appearance and behaviour of these lines has an important bearing on current theories of shock waves in the atmospheres of these stars.

The absorption lines have been shown by Preston and Paczynski to follow almost immediately after the appearance of the emission lines and have yielded valuable information concerning the surface temperatures of these variables.

In addition to the lines of hydrogen, those due to helium and several neutral and ionised metals have been found in these spectra.

Shock Wave Phenomena in RR Lyrae Stars

As has been said earlier in this chapter, the true cause of the light and spectral variations in the RR Lyrae stars is not known with absolute certainty although the theory of pulsation explains many of the available observations. We shall now go into this problem in a little more detail.

Radial velocity curves have been obtained for many of the Cluster variables although most of them have been derived from low-dispersion spectrograms. Both RR Lyrae and SU Draconis have however been investigated using spectra with a high dispersion and from these it has been found that different radial velocity curves are obtained depending upon which lines in the spectrum are used. It would appear that this is because the various lines, e.g. those due to hydrogen, the H and K lines of ionised calcium, and the various metals originate in different layers of the star's atmosphere, the lower the layer, the smaller is the amplitude of the radial velocity curve.

From the observed fact that these amplitude differences are present at all phases, it would seem that a velocity gradient exists within the stellar atmosphere. The radial velocity curve based on the metallic absorption lines is continuous and no doubling of these lines has been observed. One further relevant feature is that where doubling of the absorption lines occurs, it does so a little before maximum light and this, combined with the fact that many variables of Bailey's type **a** show emission lines of hydrogen, provides us with strong evidence of the existence of a shock wave in the upper layer of the atmosphere. It is possible that these emission lines originate in hot gas which moves outward from the star's surface behind the wave front.

Since no doubling of the lines due to metals has been observed, we may conclude that the shock wave must originate in the outermost layer of the star above that in which the metal absorption lines are formed. There is no evidence whatever to indicate that such a shock wave starts in the centre of the star where the temperature and pressure are greatest and works its way out to the surface.

There is however a good deal of observational evidence to suggest that in the case of RR Lyrae at least, the distance travelled by the shock wave is quite large implying that this star is surrounded by a very extended atmosphere of an extremely tenuous nature. Quite possibly, such a low-density atmosphere may even be supported by the movement of the shock waves as they pass through it. The discovery of the shock wave phenomenon has recently led to much speculation by astronomers on the dynamics of the atmospheres of these stars for it would seem that without some form of continuous turbulence there should be a continual infall of material due to the high gravitational field of these variables.

6. Cepheid Variables

Undoubtedly one of the most important classes of variable star from the point of view of the determination of interstellar and intergalactic distances is the Cepheids. The first to be discovered, by Goodricke in 1784, was δ Cephei itself, one of the brightest of the class. Almost a thousand are known at the present time, both in our own galaxy and in the extragalactic nebulae.

The light curves of the Cepheids are very similar to those of the RR Lyrae variables which have been described in the previous chapter. They are all characterised by a rapid rise to maximum brightness followed by a slower decline, often with a very pronounced hump on either the ascending or descending branch of the light curve. The Blazko Phenomenon is rarely, if ever, found and the light variation proceeds with a clockwork regularity. The light curve of a typical Cepheid is shown in Fig. 18.

Gaposchkin has demonstrated that as Cepheids of increasing period are considered, the irregularities in the light curve generally become more pronounced. In particular the hump on the curve grows more and more well defined and in some cases it may even coincide with the maximum producing an abnormally sharp peak to the curve. Why this should be so is not yet known and it is by no means a universal characteristic since there are Cepheids with long periods of the order of 50 to 60 days whose light curves are quite smooth. From a study of a large number of these variables it has also been found that the period distribution of the Cepheids is not uniform. The most favoured periods are clustered around 5 days, 13 days and 45 days. We shall see later how the period of a Cepheid is determined by its original position on the main sequence before it begins to evolve to the right of the Russell-Hertzsprung diagram.

Originally it was thought that the RR Lyrae stars and the Cepheids shared a common relationship and that, in spite of the well-defined gap around a period of one day, the difference between the two classes was one of degree and not of kind. This erroneous assumption was indeed responsible for many

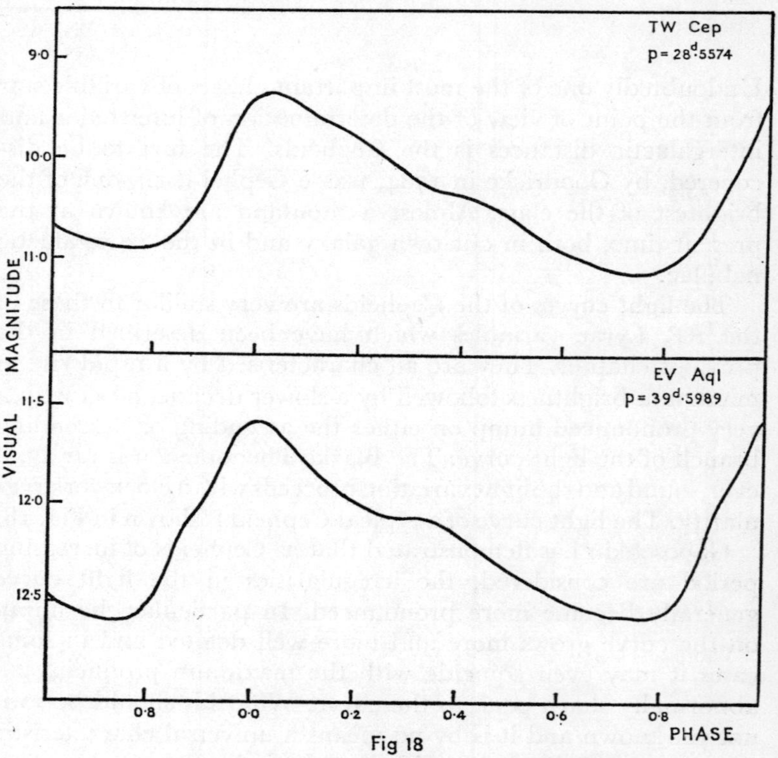

Fig. 18 – Light curves of two long period variables.

false estimates of size and distance during the early determinations of the size and shape of the galaxy and the distance to the extragalactic nebulae. Subsequent discoveries have shown that the two classes of variable are quite distinct and recognition of this fact cleared up several serious discrepancies.

As will be seen from Table 5 the great majority of these stars are of spectral type F at maximum indicating a surface temperature of the order of 7,500°K. From their spectra, in particular the Doppler shift of the absorption lines, all of the indications are that they are pulsating stars just like the RR Lyrae variables but with longer periods. One peculiarity is that although one might expect maximum brightness to correspond exactly with minimum size (when the surface temperature should be at a maximum), it actually occurs about a quarter of

Table 5
CEPHEID VARIABLES

Star	Maximum	Minimum	Period (days)	Spectrum
U Aql	6.05	6.89	7.023838	Go – G6
η Aql	3.69	4.40	7.176678	F6 – G4
SY Aur	8.80	9.50	10.144025	Go – G2
RU Cam	9.26	10.51	22.1725	Ko – R2
TU Cas	7.50	8.72	2.13930	F5 – G2
δ Cep	3.78	4.63	5.366306	F5 – G2
SU Cyg	6.37	7.08	3.845507	Fo – G1
β Dor	4.50	5.71	9.842603	F2 – F9
ζ Gem	3.70	4.09	10.153527	F7 – G3
K Pav	4.81	5.68	9.0838	F5 – G5
S Sag	5.80	7.00	8.382172	F6 – G5
W Sgr	5.00*	6.39	7.594660	F2 – G6
RV Sco	6.99	8.10	6.06130	F5 – G5
ST Tau	8.48	9.61	4.034229	F5 – G5
α UMi	2.48	2.62	3.96961	cF7v

* Photographic magnitude

a cycle later. In other words, the light curve and radial velocity curve for any particular Cepheid are out of phase by a small amount. We must remember however that not only the temperature but also the surface area of the star has a profound bearing on the total luminosity and although the temperature falls as the star expands the greater surface area more than compensates for this over a quarter of the period.

The technique for the determination of great distance by means of the period-luminosity relationship was discovered early this century by Miss Leavitt during her investigations of the numerous Cepheids found in the Magellanic Clouds.

These are two star clouds visible only from the southern hemisphere, looking to the naked eye rather like two detached portions of the Milky Way. They are, in fact, irregular nebulae closely associated with our own galaxy but not part of it.

When the periods of these Cepheids, all of which by the way are classical Cepheids with periods between 3 and 50 days, were plotted against their apparent magnitude at maximum, it was found that there was a definite relationship between these two parameters. The longer the period, the brighter was the apparent magnitude. The fact that this was found for Cepheids in the Magellanic Clouds rather than for those in our own galaxy was of great significance for since these two large star systems are small compared with their tremendous distance from us (about 80,000 light years), we may take it that all of these stars are at essentially the same distance and hence their apparent magnitudes differ from their absolute magnitudes by a constant factor. Consequently there is a relationship between period and absolute magnitude and this is known as the period-luminosity law.

Hence, if we know the period of a Cepheid and therefore its absolute magnitude, we may calculate its distance simply by means of the following equation:

$$M - m = 5 - 5 \log D$$

where M is the absolute magnitude, m the apparent magnitude and D is the distance in parsecs. A parsec is the distance of a star which subtends an angle of one second of arc of parallax. It is an even larger unit of distance than the light year being equal to 3.26 light years.

The remaining problem, of course, was to find the zero point for the period-luminosity curve, that is to establish the absolute magnitude for any one Cepheid and this proved to be an exceptionally difficult task. The nearest classical Cepheid is much too distant for the usual direct method of parallax to be used and even spectroscopic methods were of little practical value for an accurate determination. δ Cephei itself is more than 300 light years away and other members of the class are even more distant objects.

Accordingly, before the difference was recognised, the RR

Lyrae variables were used since, as mentioned earlier, these are present in large numbers in the globular clusters and their mean distances could be determined with a fair degree of accuracy by statistical methods which involve the motion of the sun towards the apex in the constellation of Lyra. These measurements all indicated a mean absolute magnitude for the RR Lyrae stars of 0.0 and using this, the distances of the classical Cepheids in the galaxy, the Magellanic Clouds and also the great spiral nebula in Andromeda, were calculated. From this work, the distance of the Andromeda spiral was found to be approximately 300,000 parsecs.

However, with the advent of the 200-inch reflector at Mount Palomar, it was discovered that whereas the photographic limit of this instrument is magnitude 24, there appeared to be no sign of any RR Lyrae variables in the nebula and from a distance of only 300,000 parsecs these should have appeared as objects of photographic magnitude 22.4. The only conclusion which could be drawn (since it was unthinkable that this nebula which is so like our own contained none of these variables) was that the nebula was more distant than had been calculated using the Cepheid scale as calibrated from the RR Lyrae stars.

Baade then turned to the long period variables which have also been identified from their characteristic light curves and typical late-type spectra in the Andromeda nebula. Here it was discovered that their mean magnitudes were about 22.4 and since astronomers have found their mean absolute magnitude from repeated and generally concordant results to be about −1.5 he was forced to the conclusion, which was then generally admitted, that the distance modulus (D) in the above equation had to be increased by a factor of 1.5. In other words, the distance to the nebula had to be doubled.

With the more recent discovery of RR Lyrae stars in the Magellanic Clouds (as opposed to the classical Cepheids which are well known), and the realisation that the Cepheids do not fall on a single line in the period-luminosity diagram but along a band which is a little more than one magnitude wide, further modifications became necessary to the original estimates of intergalactic distances showing that these should be trebled, rather than merely doubled. This places the Andromeda spiral at a distance of 2,200,000 light years and the Magellanic Clouds at about 170,000 light years as against the 80,000 light years mentioned earlier.

The above findings have since been confirmed by a study of novae and blue giant stars in the nearer nebulae. One other fact which was known before the necessary modifications were made to these distances and which made many astronomers of the time suspicious of the original measurements was that the earlier adopted dimensions of the extragalactic nebulae showed that our own was a veritable giant, far larger than even the biggest among them. From the foregoing discussion it may be seen that the fact the Cepheids are quite distinct from the RR Lyrae variables in spite of the other marked similarities between them, has led to far-reaching changes in our knowledge of the dimensions of the universe as a whole.

Evolution of the Cepheids

We must now consider in more detail the evolution of the classical Cepheids compared with that of the RR Lyrae stars. In the last chapter we saw that once a core of helium is formed in the centre of a dwarf star very similar to the sun, it evolves to the right of the main sequence following a path as shown in Fig. 14. That this is true for stars with masses and luminosities like the sun we know from a study of the stars in the individual globular clusters where we can observe such stars at all points along this curve of evolution.

Now we must examine the analogous case where stars lying above the dwarfs on the main sequence begin to evolve and see whether, given enough time, they too will follow a similar track to the yellow dwarfs once helium production begins in the core. An examination of the Russell-Hertzsprung diagram for other clusters and also our own galaxy shows that this is indeed so in some cases but not in others. However, it is true for stars just above the region of the yellow dwarfs on the main sequence and by analogy with the RR Lyrae variables we would expect to find a point in very much the same position along the track of evolution where oscillations occur. (Fig. 14.)

Here, however, instead of a point there is a vertical zone in which pulsation occurs and depending upon the star's initial position along the main sequence, it will cross this zone of instability at different points. (Fig. 19.)

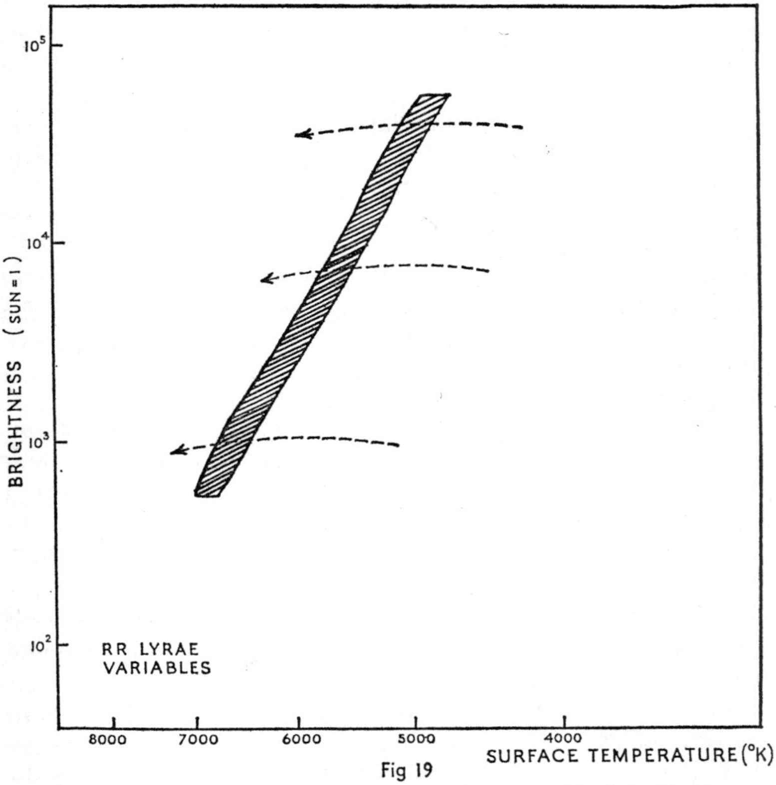

Fig. 19 – The Cepheid instability zone (*courtesy of Prof. F. Hoyle*).

These particular oscillating stars are the Cepheids and from observation we know that the entire span of the Cepheid instability zone is covered by stars having an initial main sequence brightness between 10 and 300 times that of the sun. Not only this, but the period of the pulsation depends upon the point where the evolutionary track crosses the Cepheid zone, varying from one day at the lower end to approximately seventy-five days at the higher. This then explains the period-luminosity relationship found for these variables.

Clearly if the period of a particular Cepheid is known we can tell at which point along the Cepheid zone it lies and the intrinsic brightness may be read off directly. Since the most luminous of the Cepheids is more than 350 times brighter than the RR Lyrae stars they can clearly be employed to estimate far greater distances, up to about 12,000,000 light years. There can be no doubt that these variables have provided astronomers with one of the most powerful tools for giving us a reasonably accurate picture of the universe.

Population II Cepheids

We have now reached the point where we must consider one of the enigmas associated with the Cepheid variables, that of the curious absence of these stars from the globular clusters. We have already seen that these large groups of stars situated on the periphery of the galaxy contain large numbers of the RR Lyrae stars but few, if any, classical Cepheids with periods greater than one day. Quite evidently, the stellar population of the globular clusters appears to bear no comparison to that in the neighbourhood of the sun.

Earlier in this chapter we saw the great difficulties which arose when astronomers attempted to measure distances on the assumption that irrespective of period, all RR Lyrae and Cepheid variables followed the same period-luminosity law and it was not until the distinction between the two types of stars was recognised that concordant results were obtained. It is now known, mainly from researches due to Baade, that there are two different types of stars within the galaxy; those within the spiral arms which are termed Population I stars and those in the nucleus of the galaxy or in the halo (which includes the globular clusters) which are known as Population II stars. This division by Baade into the two stellar populations has resolved many of the earlier difficulties and anomalies. It is generally believed that the stars of Population II were formed out of the primal matter of the glaxy and consequently they contain a lower percentage of metals in their composition than the later stars of Population I. This is mainly a result of the formation of the heavier elements within the cores of the hypermassive Population II stars by nuclear synthesis and contraction. Theory has shown that such stars may often experience a supernova outburst, the explosion scattering the heavy elements throughout interstellar space so that the Population I stars which were formed at a later date did so from an interstellar gas already relatively rich in metals.

We shall now see how the Cepheids fit into this concept of two distinct stellar populations. Quite obviously, the large majority of the RR Lyrae variables must be Population II stars since they are found in large numbers within the globular clusters. But what of the Cepheids? Are they predominantly of Population I or do they belong to both types but with characteristics so similar that it is impossible to distinguish between them?

Studies of the many hundreds of Cepheids within the Magellanic Clouds on which the period-luminosity law has been based, show that they all appear to obey the same law, that their light curves follow a similar pattern and the period-amplitude relationship holds for them all. Since these variables are found in the galactic planes of the two Magellanic Clouds it follows that they are Population I stars.

W Virginis Variables

To gain further insight into this fascinating problem we must return to the Cepheids within our own galaxy. Here we find that two types of Cepheids may be distinguished; those typified by δ Cephei which belong to the galactic plane and those whose prototype is the tenth magnitude star W Virginis which are members of the galactic halo. Very significant physical differences have been found for these two kinds of variable.

Whereas the light curves of the Population I Cepheids are essentially constant from one cycle to the next, quite appreciable changes have been observed in those of the Population II Cepheids. For example, the two variables U Saggittarii and S Normae which are both in galactic clusters, M 25 and NGC 6087 respectively, show a pronounced hump on the descending branch of the light curve which varies perceptibly from one phase to the next. A variable hump is also present in the radial velocity curves for these two stars.

Observation also shows that the colours of the two types of variable are not the same and it is very probable that their composition is also different, those of Population I containing a higher proportion of metals. Most significant of all perhaps is that the period-luminosity curve for the Population II Cepheids lies parallel to that of the Population I variables but for any given period is about 1.5 magnitudes fainter.

This immediately raises a very interesting point concerning the mystery of the missing Population II Cepheids in the Magellanic Clouds. Since there does appear to be this definite difference in the period-luminosity curve for the two populations we would expect to find evidence of Cepheid variables of Population II merely from a study of their period-luminosity curves but as it turns out, of more than 600 such variables in the Magellanic Clouds, only two have been found more than one magnitude below the mean curve. Are we then led to the inescapable conclusion that there are no Cepheids of Population II at all in these particular nebulae? At first sight this would indeed seem to be the case.

At this point it is important to realise the nature of the two Magellanic Clouds. They are good examples of the irregular nebulae being fairly open systems and it would be only logical to assume that in any initial survey, the Cepheids which have been recognised so far will be those which lie in or near the galactic plane since it is here, more than anywhere else, that such variables will be more readily discernible and their type of light variation recognised.

Although not completely absent, it is quite possible that the Population II Cepheids are abnormally scarce in the Magellanic Clouds and if this is so then they will be found, if anywhere, in regions which we know to be rich in Population II stars, namely in the globular clusters around these two nebulae. Such a search has recently been carried out with the unexpected result that there is virtually no difference in the period-luminosity curves for the two populations of Cepheid variables.

It would now appear that we have solved one mystery only to be presented with another. The missing Population II

Cepheids have been found among the globular clusters of the Magellanic Clouds but the period-luminosity relationship is not that which has been discovered for such stars within our own galaxy. Perhaps the most plausible answer to this at present is that suggested by Hodge and Wright who have shown that when the period-luminosity curves for Population II Cepheids in different galactic globular clusters are compared they form a series of curves which are approximately parallel but differ in level by more than a magnitude. This difference has been interpreted as a measure of the abundance of metals in the various clusters. We have already seen that the Population II stars are, in general, relatively deficient in metals compared with those of Population I. From other considerations such as the colour diagrams of these various clusters, it is found that those in the Magellanic Clouds are richer in metals than others within our own galaxy. In this respect therefore, these variables are closer in composition to the Population I Cepheids.

The Spectra of Cepheid Variables

It is not an easy matter to draw any clear-cut distinction between the spectra of Populations I and II Cepheids. Certainly transient emission lines of ionised calcium are a characteristic feature of the stars of Population I, lines which have been noted in only two Population II Cepheids, W Virginis and RU Camelopardalis. The latter variable is not a typical Cepheid by any means, having a very late-type spectrum varying from Ko to R2. Indeed, there is even evidence that within the past year or so this star has ceased its light variation! Primarily, the spectra of Population II Cepheids are characterised by emission lines of hydrogen which behave in a similar manner to those of the RR Lyrae stars, appearing about midway between minimum and maximum brightness. Some Cepheids have been found in which no hydrogen emission appears in the spectrum but for those of Population II in which it is present there exists a definite relationship between the period of the star and the fraction of the period over which it appears; the longer the period, the greater is the fraction during which the emission lasts.

Another important straight-line logarithmic relationship has been found by Kraft between the period and the length of time during which the calcium emission lines are present and it is possible to extrapolate to periods even longer than those of the Cepheids when we find that the calcium emission of certain long period variables also satisfies this relationship. Here then we have spectral evidence that the mechanism responsible for the emission is the same in both types of variable.

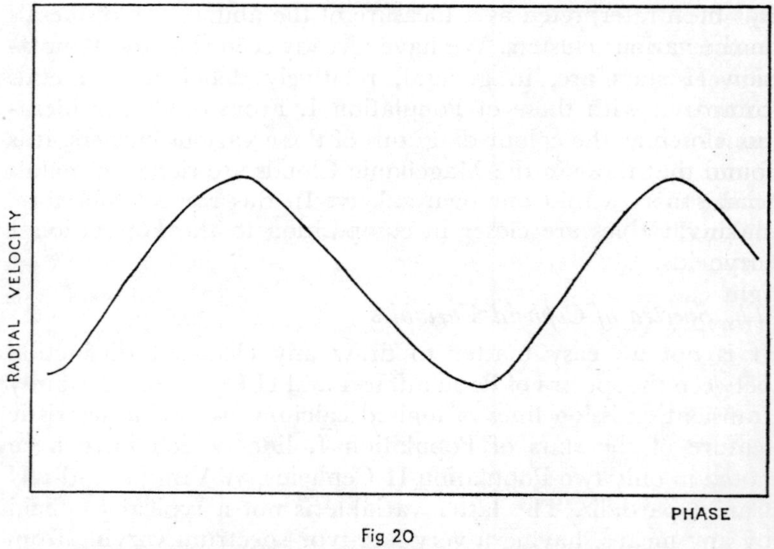

Fig. 20 – Radial velocity curve of a typical Cepheid variable.

Radial Velocities and Shock Wave Phenomena

The radial velocity curves of the Cepheid variables are very similar to those of the RR Lyrae stars as we might expect. (Fig. 20.) Accordingly we may postulate the same kind of mechanism as being responsible for the light variations, namely a pulsation of the outer layers of these stars.

The emission lines and other spectral peculiarities in the spectra of the Cepheids of which BL Herculis and W Virginis have been most thoroughly investigated are in agreement with the development of a shock wave in the atmosphere possessing sufficient energy to ionise much of the gas in addition to compressing it and overcoming the gravitational pull of the star as the wave moves outward. As well as the shock wave moving outward, there will be material falling back in towards the star and in support of the shock wave hypothesis it has been found that this infalling gas is always outside the shock wave front moving outward.

The Period-Colour Relationship

By means of interference-filter measurements Kraft has demonstrated that there is an extremely close relationship between the period of a Cepheid and its mean colour, especially for those variables which have large amplitudes. A similar, but less close relationship exists between period and the intrinsic colour as measured at maximum light, the wider scatter found by this method being presumably due to inherent difficulties in making these measurements at only one point along the light curve. As the mean colour takes into account the whole of the light curve it will accordingly be less subject to observational errors than observations made only around maximum.

7. β Canis Majoris Variables

One particular class of variable star usually associated with the Cepheids is that whose prototype is the magnificent second magnitude star β Canis Majoris. The amplitudes of these variables are all extremely small, rarely exceeding 0.1 magnitude and for this reason very sensitive photoelectric equipment is necessary to follow their light variations. Owing to this very small change in brightness, these stars have so far been discovered only among the brighter stars although from the total number found it is evident that they must exist in great numbers throughout the galaxy. A list of the better known variables of this class is given below in Table 6.

Table 6

β CANIS MAJORIS VARIABLES

Star	Maximum	Minimum	Period (days)	Spectrum
γ Boo	3.20	3.25	0.2903137	A7
α^2 CVn	3.00	3.10	5.470	A0p
β CMa	2.02	2.08	0.250031	B1
β Cep	3.32	3.37	0.1904851	B1s
δ Cet	4.00	4.03	0.1611366	B2
DD Lac	5.00	5.13	0.1930890	B2n
BW Vul	6.21	6.38	0.201030	B1s

All of the above magnitudes are photoelectric

As will be seen from Table 6, these are all very hot, white stars and all have very short periods with the single exception of α^2 Canum Venaticorum and there is some doubt as to whether or not this variable is actually a member of this particular class. The strangest feature of these stars is that after varying in quite a regular manner for some years, the light fluctuations die out and for a long period, often several years, the light variations begin again, still following the earlier, well-defined period. This peculiarity is especially noticeable in the case of γ Boötis.

There are also odd changes in the spectra of these variables which often show irregularities of both shape and amplitude in their light curves although generally the radial velocities keep in synchronism with the light curve changes.

Fig. 21 shows the light curve of a typical β Canis Majoris variable.

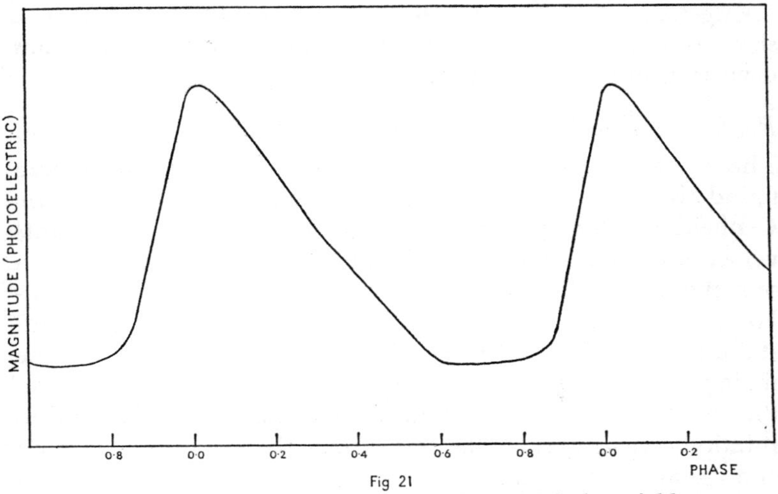

Fig. 21 – Light curve of a typical β Canis Majoris variable.

On the Russell-Hertzsprung diagram these stars form a clearly defined sequence from spectral type A7 to B2 and it has been known for some time now that their periods are closely related to the star's position along this sequence. It also seems likely that the brighter members of this class have had time in which to evolve away to the right of the main sequence whereas the less luminous members have evolved only slightly.

Spectra of β Canis Majoris Stars

Being hot stars with surface temperatures in the range 8,000 to 16,000°K, the predominant absorption lines in the spectra of these variables are those of hydrogen, neutral helium, singly ionised carbon, nitrogen and oxygen together with those of ionised metals. The absorption lines due to ionised calcium originate in the stars themselves and not, as is the case with the more distant B-type stars, in interstellar space since they are found to oscillate in the same way as the other absorption lines and show no systematic differences.

Since the absorption lines are narrow there seems no evidence from the spectra for any motion of the gas in the outer layers of these variables as would be the case if any expansion were taking place. On the other hand, the light variations tend to show that there are faint but more or less stable pulsations occurring in these outermost layers.

Rotational Velocities and Pulsational Instability

The sharp, narrow absorption lines mentioned above and found also in the spectra of other pulsating variables such as the Cepheids and long period stars are indicative of a low rotational velocity. Earlier, we saw that the Cepheids were originally main sequence stars which evolved towards the right of the Russell-Hertzsprung diagram eventually crossing the Cepheid instability zone. Like all stars which evolve to the right of the main sequence, expansion occurs with a concurrent lowering of the surface temperature and also a marked reduction in rotational velocity since they must obey the law of conservation of angular momentum.

In sharp contrast to the other types of pulsating variables, however, the β Canis Majoris stars have high surface temperatures and are almost certainly young stars, found in the region of the Russell-Hertzsprung diagram occupied by stars having predominantly large rotational velocities. The question therefore arises – are these variables pulsating stars because of this low velocity of rotation? In other words is there a particular rotational velocity for these bright stars above which they do not pulsate and below which they fall into the β Canis Majoris class?

The available evidence so far gathered suggests that this may indeed be the case. All of the β Canis Majoris variables have rotational velocities less than 40 kilometres per second whereas the mean rotational velocity for the non-variable stars of the same spectral class is far higher, of the order of 150 kilometres per second. We are thus led to the conclusion that the axial rotation of a star in this particular region of the Russell-Hertzsprung diagram is critically linked with pulsational instability which produces the characteristic light curve of these variables. Theoretical approaches to the problem have indicated that all of these hot B-type stars entering this region will be dynamically unstable and provided they have a sufficiently low velocity of rotation they will pulsate. As the rotational velocity increases, however, beat phenomena enter into the pulsation until above a rotational velocity of about fifty kilometres per second, the rotation is now rapid enough to suppress the pulsation and the star, although having the same spectral type, will be non-variable in brightness.

Our understanding of the dynamical processes within these stars is still at a very rudimentary stage and it is not possible to give a precise picture for the reasons why pulsation takes place although it has been suggested that helium burning in the core and the consequent changes taking place at the centre of the star are the basic cause of instability.

Radial Velocity of β Canis Majoris Variables

Radial velocity curves have been derived for most of these stars and have shed some light on the conditions prevailing within the outer layers of these variables. In every case, there is a phase lag in the curves as derived from the hydrogen or helium absorption lines relative to those determined from the other absorption lines due to the ionised metals. Like the other pulsating variables previously discussed, this may be interpreted as due to a shock wave moving outward through the atmosphere of the star.

It will be appreciated that the results obtained are based on only a small number of stars since the known β Canis Majoris variables are by no means as numerous as the RR Lyrae or Cepheid variables; nevertheless there does appear to be some correlation between the amplitude of the radial velocity curve and the depth within the stellar atmosphere at which the particular absorption lines originate.

8. δ Scuti Variables

Like the β Canis Majoris stars, the δ Scuti variables are pulsating stars of very short period characterised by quite small light and radial velocity amplitudes, frequently exhibiting beat

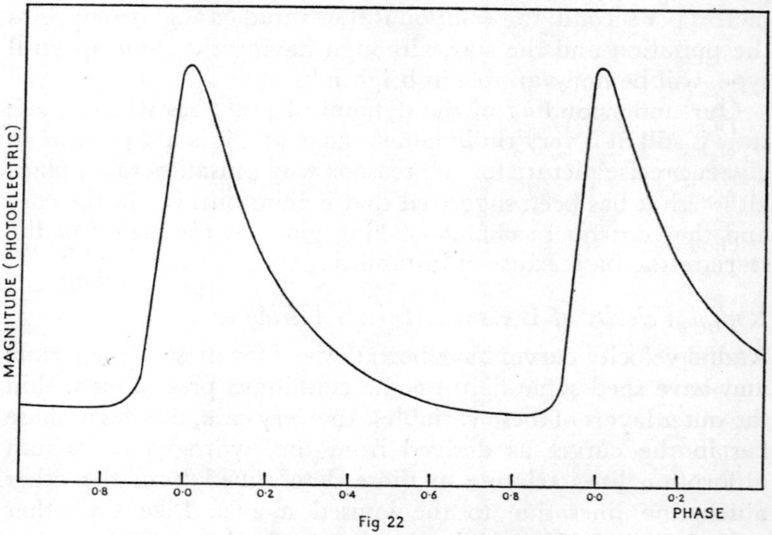

Fig. 22 – Light curve of a typical δ Scuti variable.

phenomena in both light and radial velocity curves. Very few members of this small class are known with certainty and all have either A or F type spectra. One major difference between these stars and other types of pulsating variable is the abnormal phase anomalies in the relationship between their light and radial velocity curves.

For both ρ Puppis and δ Delphini the maximum in the light curve occurs about midway along the ascending branch of the radial velocity curve while for DQ Cephei it is halfway down the descending branch. More recently, Fitch has demonstrated that in the case of δ Scuti, maximum brilliance precedes minimum radial velocity by 0.07 period.

As will readily be seen from Fig. 22, the light curve of a typical δ Scuti variable closely resembles that of an RR Lyrae star of Bailey's class **c**.

Table 7 lists the data on five of these variables which have at present been assigned to this class.

Table 7
δ SCUTI VARIABLES

Star	Maximum	Minimum	Period (days)	Spectrum
CC And	8.45	8.69	0.1249078	F3
DQ Cep	7.40*	7.48	0.0788650	F1
δ Del	4.50	4.56	0.13505	F2
ρ Pup	2.68	2.78	0.1410	F6
δ Sct	4.90*	5.19	0.193770	F3**

* Photographic magnitude
** The spectrum of this star has been given as Type A9

The luminosity class assigned to these stars suggests quite a large spread in absolute magnitude from −2.0 to +2.2 and they would appear to be situated about one magnitude above the main sequence stars and very close to the RR Lyrae variables which they closely resemble.

Two of these variables, δ Delphini and ρ Puppis, both show a deficiency of calcium in their spectra, so much so that the calcium K-line intensity in the former is similar to that of an A7 star although the other lines in the spectrum are in agreement with type F2.

Provided that the normal relationships for pulsating stars with a high central condensation hold for the δ Scuti variables, the masses have been estimated as ranging from 0.6 that of the sun for δ Scuti to 3.5 solar masses for DQ Cephei. There is therefore also a wide spread of masses among these variables. The average mass of 1.9 times that of the sun would suggest that they are A type stars originally on the main sequence which have evolved to the right. We must also take into account Eggen's view that they are very similar to the RR Lyrae variables which indicates that they are evolving to the left of the main sequence.

The β Canis Majoris, or dwarf Cepheids, have been described in the previous chapter and it is interesting to note that those with periods between 0.10 and 0.15 day lie immediately to the left of the δ Scuti variables on the Russell-Hertzsprung diagram and it is perhaps tempting to postulate that there may be a possible generic connection between the two. There is, however, no real evidence to support this theory. Indeed, the brightness amplitudes and also the phase relationships between light and radial velocity curves are so different that it is difficult to see how these two classes of stars, so different in many ways, could be related.

Another piece of evidence to which we may look for some idea of the direction in which these stars have evolved is their velocity of rotation. Since they are quite bright stars, this may be readily determined from the Doppler shift of the spectral lines and we find that the average rotational velocity thus measured for these stars is approximately 35 kilometres per second. This fits the suggestion that they have evolved to the right of the main sequence into their present position.

Returning to Eggen's view, it is noticeable that unlike the Cepheids or RR Lyrae stars, the δ Scuti variables have very small colour indices and their spectral type remains essentially constant throughout the whole of the light cycle.

Many attempts have been made to explain the light variations of δ Scuti for which several series of extremely accurate photoelectric observations have been provided by Fath. From his own observations and those of Nikonov, this observer concluded that there were three incommensurable periods for this star, the mean light curve being very nearly sinusoidal within each period. Owing to uncertainties present in the mean light curves, Fath was not able to distinguish a resonance term from the first overtone.

Since we know that resonance phenomena do occur in several similar stars, particularly RR Lyrae and CC Andromedae, it is only natural to assume the same for δ Scuti. A detailed analysis by Fitch has shown that the light variations of this star may be explained on the assumption of a primary period of 0.19377 days, together with the second overtone of 0.09513 days and a resonance term of 0.18687 day. From a study of the radial velocity curves, the resonance term is shown to be more significant than the second overtone.

9. Magnetic Variables

The magnetic variables belong to a class of peculiar A type stars and although only a handful are known, these are mostly bright, naked-eye objects. Their variation in light is however extremely small, seldom exceeding 0.1 magnitude and as a result the light changes can only be followed accurately by photoelectric techniques. (Fig. 23.)

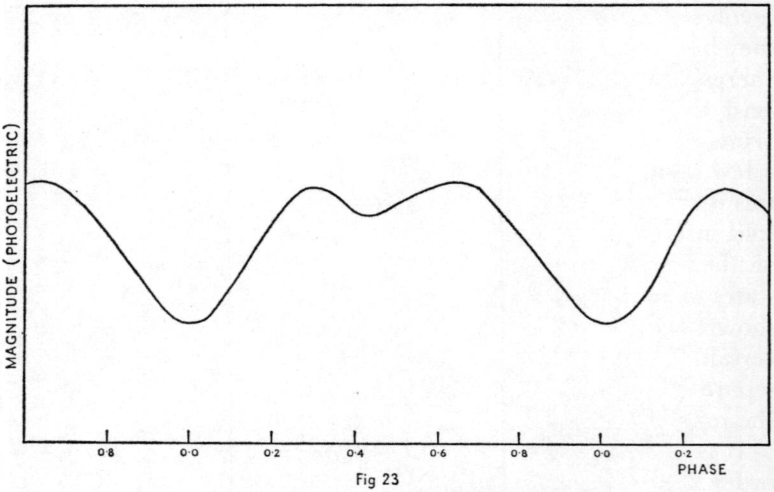

Fig. 23 – Light curve of a typical magnetic variable.

The light curves of these stars bear many similarities to those of the β Canis Majoris variables discussed in an earlier chapter, so much so that the type star of this small group, α^2 Canum Venaticorum, has long been considered as a β Canis Majoris star. (Table 6.)

If, therefore, there is this very great similarity in their light variations, what is the difference between these two classes? The answer is to be found, not in their light curves, but in their spectra. The β Canis Majoris stars are giants of spectral classes B1 to B3, while the magnetic variables are all of the succeeding class A. Apart from this, the most significant difference lies in the behaviour of the spectral lines. The spectra of the magnetic stars all show intense lines of strontium, silicon, chromium and the rare-earth metals, the position and intensity of which vary with the light period. A similar effect is found too for the lines of neutral hydrogen. More important still, the spectra show that all of these stars possess very powerful magnetic fields, the proof of this being found in the Zeeman Effect which was discovered in 1896.

Briefly, when an atom of a radiating or absorbing gas is acted upon by a magnetic field, the orbits in which the electrons revolve around the nucleus are orientated differently to those they have in the absence of such a field, resulting in a change of energy for such an atom. We now know that there does not exist an infinite number of spatial orientations which these orbits may take up, the number actually being limited to only a few by what is known as spatial quantisation. The result of this is that the emission or absorption lines in the spectrum are split into two or more closely-spaced components on either side of the original line because of the slightly different energy states now possessed by the various atoms. Experiment has shown that provided the magnetic field does not exceed a certain critical value (about six thousand gauss), the degree of separation of the lines is directly proportional to the strength of the magnetic field.

It is, of course, necessary to have high-dispersion spectra in order to observe the Zeeman Effect in stars, even more so to actually measure the strength of the field since the splitting of the lines in the spectrum is exceedingly small. Caution is also necessary in the evaluation of the magnetic field strength, or even in making the assumption that it really exists for a certain star, since there are other external effects which also produce line splitting. We have already encountered one such cause, namely the doubling of the absorption lines due either to the presence of two stars physically connected and revolving around a common centre of gravity, or to the axial rotation of a single star.

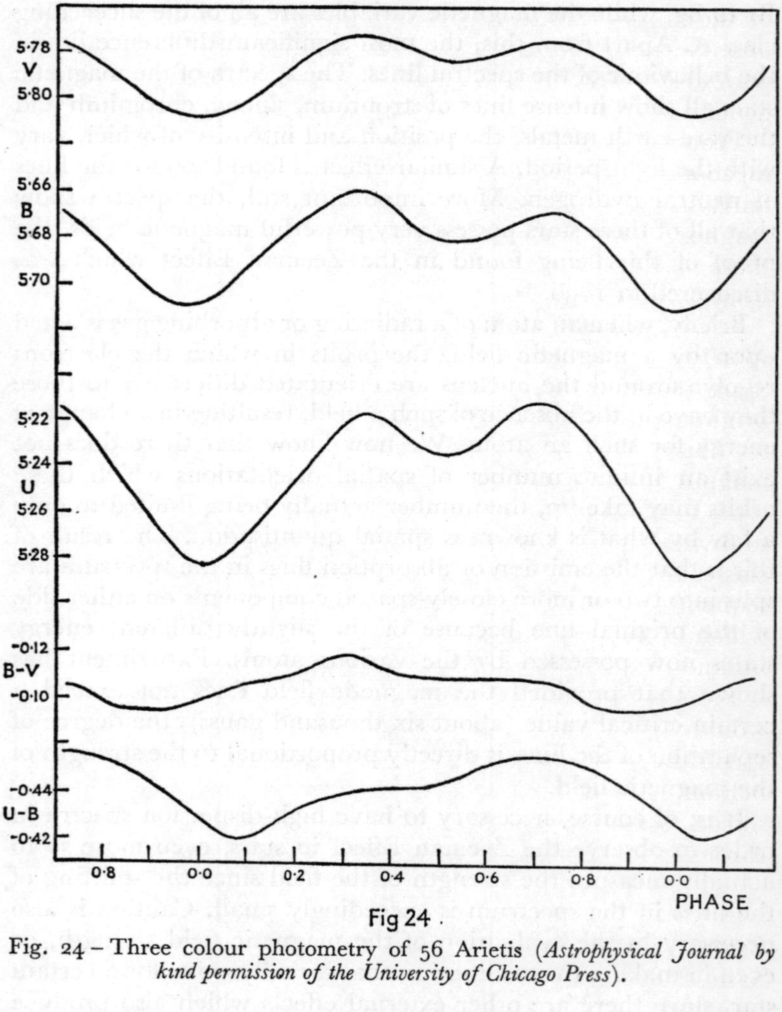

Fig 24

Fig. 24 – Three colour photometry of 56 Arietis (*Astrophysical Journal by kind permission of the University of Chicago Press*).

Light Curves of the Magnetic Variables

A typical light curve is shown in Fig. 24 from which it will be seen that particularly in the blue and ultra-violet there are marked asymmetries in the curves and that the times of minimum do not coincide being earliest in the blue and latest in the ultra-violet. This particular variable, 56 Arietis, is one of the most thoroughly observed of all these stars and from observations made over several cycles, it appears that the period of 0.7278925 day is reasonably constant as is also the general shape of the light curve.

From this it would seem logical to assume that whatever the mechanism responsible for the light variations may be, it is fairly stable and is possibly connected with the presence of the magnetic field. The periods of these variables are longer than those of the apparently related β Canis Majoris class, ranging from 0.7 to as much as twenty-five days.

Spectra of the Magnetic Variables

Since these stars are mostly naked-eye variables, several being brighter than seventh magnitude, it is a relatively easy matter to obtain high dispersion spectrograms and measure the magnitude of the Zeeman Effect. All of these stars have spectral types between B8p and F0p and it is possible – and convenient – to divide them into two groups; the early-type stars of spectral classes B8p to A2p which are found to lie very close to the main sequence and those of the later type A and the F0p variables which lie below the main sequence stars and which, from their luminosities, are clearly giants.

From the colours of the early-type magnetic variables, they would appear to be normal dwarf stars and are classified spectrally by the gradually decreasing intensity of the neutral helium lines relative to those of singly ionised magnesium or silicon which have been shown to remain constant in strength throughout the sequence.

Cause of the Light and Spectral Variations

Before we can reach any definite conclusions concerning the basic cause of the light and spectral changes occurring in these peculiar A-type stars, there are several fundamental questions which remain to be answered. Firstly although quite a large number of these peculiar A stars are known, only a relatively small number have been shown to be either variable in light or to possess a magnetic field. It is therefore pertinent to ask: Are they all variable with magnetic fields?

We must recognise that even with very accurate photoelectric measurements there is a limit beyond which no light change can be readily detected and although Provin has shown that several stars are apparently constant in their brightness, it is possible that small fluctuations of the order of 0.01 magnitude may occur. At present therefore it is not possible to state with absolute certainty that none of these stars is variable. A similar difficulty exists when we come to consider the presence of magnetic fields in these variables. Only in those cases where the spectral lines are sufficiently narrow to allow of measurements of the Zeeman Effect can we give any answer to this question. Where it is possible to make such measurements, we find that all of these stars do have magnetic fields. There are, however, several in which the lines are so broad and diffuse that it is impossible to make any measurements at all.

We do however have sufficient data to proceed to the next question that springs to mind: Can we find any correlation between the light variations and the strength of the magnetic field for these stars? Here we must assume a mean magnetic field intensity for any given star since the temporal character of these fields is either of a periodically varying nature or in certain cases totally irregular. At the present time, much of the evidence is against such a correlation. For example, the star with the strongest magnetic field so far discovered has only a small light amplitude and even when we attempt to find some relation between the changes in the magnetic field and the corresponding light fluctuations we are no better off, any correlation being of an extremely tenuous nature.

The whole question of the origin of these magnetic fields and their nature is one which is receiving a great deal of attention from astronomers at the present time. There is some evidence from other fields of stellar astronomy that every rotating body of the size of the average star will produce its own magnetic field.

10. Long Period Variables

So far we have been discussing variable stars which are relatively regular in their general behaviour. The Cepheids for example perform their light variations with an almost clockwork regularity and both the β Canis Majoris and δ Scuti stars are only slightly less regular. We now, however, come to a large and very important class known as the long period or Mira variables whose periods are an extension of those of the classical Cepheids. Now there are relatively few Cepheids having periods in excess of fifty days; DN Arae with a period of eighty-two days possibly holding the record in this respect although owing to its faintness there is necessarily some doubt as to whether it is indeed a member of this class. The long period variables on the other hand have periods between 80 and 1,000 days with a maximum around 255 days and unlike the previous classes of stars both their periods and general behaviour are far from regular.

The first star of this class to be discovered and that from which they take their name is Mira, classified as *o* Ceti by Bayer who catalogued it as a fifth magnitude star at the beginning of the seventeenth century, being ignorant of its variable nature although it had been seen by Fabricius in 1596 as a star of the second magnitude. It was left to Holwarda to supply the proof of its true nature as a variable star and in 1638 he was able to show that it varied between second and tenth magnitude in approximately 330 days.

Mira has been assiduously observed for more than a century and we now know a great deal about its light and spectral characteristics. The period between successive maxima is by no means constant but varies between 320 and 370 days and so far, in spite of long and detailed statistical analyses of its light curve, it has proved impossible to discover any regularity at all in these wide variations of period about the mean of 330 days. For this reason it is necessary to follow the light changes of Mira, and the other members of this class, continuously since there is no way at all of predicting the dates of one maximum from those which have preceded it with an accuracy exceeding 15 to 20 days. The task of keeping the long period variables under constant observation has long been the province of the amateur astronomer since their large amplitudes and reasonably gradual variation in brightness make them ideal objects for a programme of communal observation. To date more than 1,500,000 observations of long period variables have been accumulated by the members of the Variable Star Section of the British Astronomical Association and the American Association of Variable Star Observers providing us with an extremely valuable record of the past behaviour of many of these stars.

Not only the periods are subject to these wide and unpredictable variations but also the amplitude and shape of the individual light cycles. At maximum, Mira can reach anything between second and fifth magnitude and on one occasion, in 1779, it even attained first magnitude being equal to Aldebaran in brightness. At the maximum of 1924 and 1935 on the other hand it was barely visible to the naked eye. Individual minima are also variable although not to the same extent. On occasions it has only been as faint as magnitude 8.0 while at other times it has sunk as low as 10.2 magnitude.

During the eighteenth and nineteenth centuries most of the discoveries of long period variables were made by amateurs who concentrated their attentions on the red stars since it was generally believed that being cool objects there was more likelihood of finding variability in such stars than in the white or yellow stars (a supposition we now know to be true as far as the long period variables are concerned). Mention must be made of Baxendell who discovered among many others S Aquilae (1863) and U Boötis (1880); and Pogson whose discoveries include R Cygni (1852) and in the following year R Cassiopeiae, R and S Ursae Majoris. (See Table 8.)

More than a thousand long period variables have so far been discovered in the galaxy and mention has already been made of those in the nearer extragalactic nebulae which have proved useful in the determination of intergalactic distances. In contrast to the Cepheids, the long period variables do not obey the

Table 8
LONG PERIOD VARIABLES

Star	Maximum	Minimum	Period** (days)	Spectrum
R And	5.0	15.3	408.87	Se
T Aps	8.4	14.8	260.83	M3e
R Aqr	6.7	11.6	386.86	M7e+P+O
W Aql	7.2	14.0	489.28	Se
U Ara	9.2*	15.4	225.05	M4e
Z Car	9.6	15.1	382.64	M6e
S Cas	6.2	15.3	611.83	Se
T Cen	5.2	10.0	90.65	Moe
o Cet	1.0	10.2	331.48	M6e
V CrB	6.8	12.4	357.64	Ne
R Cyg	5.9	14.6	425.44	Se
χ Cyg	2.3	14.3	406.66	Mpe
R Leo	4.4	11.6	313.13	M8e
T Oct	8.4	13.6	218.46	M2e
R Ori	8.5	13.4	378.52	Se
R Psc	7.0	14.8	343.69	M4e
R Sgr	6.6	13.3	268.83	M5e
RR Sco	5.0	12.2	279.45	M6e
R Tuc	8.7	15.2	285.84	M5e
R UMa	6.2	13.6	301.21	M4e
S UMa	7.1	12.9	225.62	Se
W Vel	9.5*	15.0	393.45	M7e
R Vir	6.2	12.6	145.35	M4e
S Vol	8.5	13.7	395.95	Moe
RU Vul	8.5	11.5	155.50	M3e

* Photographic magnitude
** Mean period

period-luminosity law. Since the period, amplitude and general shape of the light curves of these stars are continuously changing, it is impossible to assign a true period to any of them. All that can be done is to give a mean period which is the average taken over a great many years.

Light Characteristics of Long Period Variables

The main forms taken by the light curves of the long period variables are now well known from the many thousands of visual observations made over the past century and a half and the mean periods and extreme amplitudes are known for the large majority of them. Mira itself proves to be one of the more

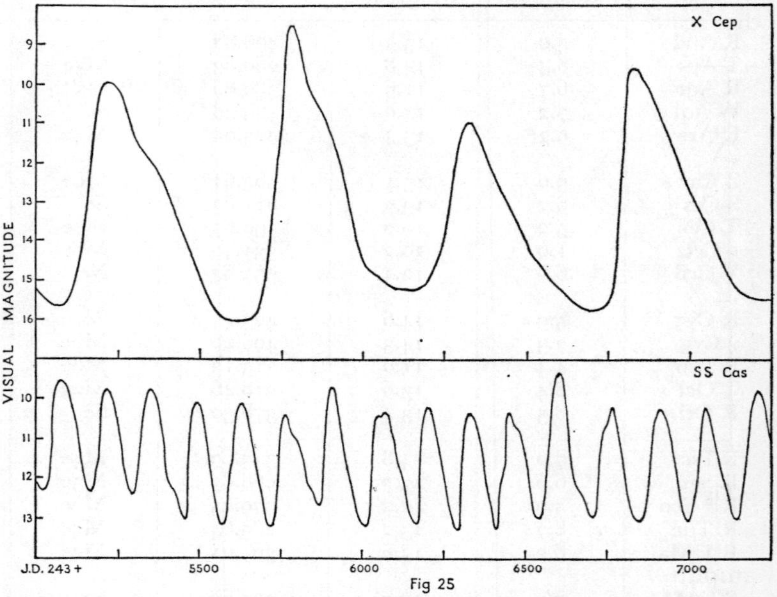

Fig. 25 – Light curves of two typical long period variables (*Courtesy of the British Astronomical Association*).

regular stars of this class. The light variations of most of them show more marked irregularities. (Fig. 25.) In only a small number of long period variables are the light curves found to be even approximately smooth; far more often they are distorted by a pronounced hump which may be either on the ascending or the descending branch of the curve. The position of this hump is by no means constant from one cycle to the next and in certain periods it may be completely absent. There are even one or two stars in which this distortion is so marked and well defined that it produces what appears to be a secondary minimum (Fig. 26) giving the light curve an appearance very reminiscent of the RV Tauri variables mentioned in a later chapter.

As far as the amplitudes of these stars are concerned, here again we find a very wide variation. Some long period variables, e.g. Y Arietis, Z Aurigae and R Ursae Minoris have amplitudes of only one or two magnitudes while at the other end of the scale are stars like R Andromedae and χ Cygni with

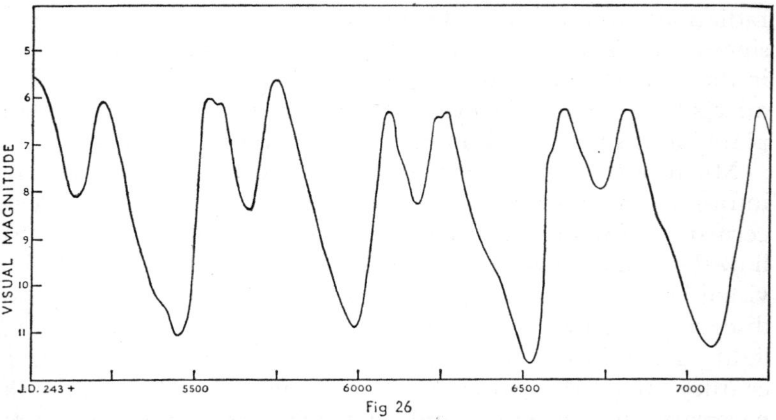

Fig. 26 – Light curve of R Centauri (*Courtesy of the American Association of Variable Star Observers*).

amplitudes of at least ten magnitudes and resembling the novae in this respect. Some astronomers have arbitrarily divided this class of stars into two groups, the long period variables with amplitudes of four magnitudes or less, and the Mira type variables in which the light variation is in excess of five magnitudes. There is however very little, if any, physical difference between the stars in these two groups and it is more a method of convenience than one representing a real distinction between the stars themselves.

Other Radiation from the Long Period Variables

The visible light we receive from the stars is not, of course, the only kind of radiation which they emit. In the whole range of the electromagnetic spectrum, the visible region is really extremely small. Extending beyond the red end lies the infra-red which, although invisible to the eye, produces a heating effect as discovered in 1800 by Herschel who used a thermometer with a blackened bulb. In 1801, Ritter showed that there were similar invisible radiations beyond the blue end of the visible spectrum which had the effect of reducing silver chloride to metallic silver (as occurs in the ordinary photographic process), these being the ultra-violet radiations. At first it was thought that the infra-red, visible and ultra-violet radiations were of three distinct kinds but this view has long since been discarded. We now know that the difference lies only in their wavelength and the fact that one particular wavelength affects the retina of the eye and another only the photographic plate implies no special properties of the radiation itself.

Much of the total energy of the stars comes through in the infra-red and ultra-violet and where variable stars are concerned it is important to be able to measure this and also find how the radiation at these different wavelengths varies with the visual light cycle. During the two centuries or so following the discovery of Mira, astronomers had to rely entirely on visual light curves since it was not until the beginning of the present century that the specialised instruments and techniques necessary for detecting and measuring the other types of radiation were first developed and then adapted for use in astronomical investigation. One of the major difficulties encountered in stellar work as opposed to solar investigation is the very small size of the image produced in the telescope but recent improvements have now overcome this problem.

One of the instruments which has made valuable contributions to our knowledge of variable stars is the vacuum thermocouple. The thermoelectric effect on which it is based was known to physicists for several decades before its use in astronomy. Essentially the thermocouple consists of two wires made of dissimilar metals welded end to end and forming part of an electric circuit. Now if one of the two junctions is made warmer than the other an electric current will flow in the circuit which may be accurately measured by means of a sensitive galvanometer, the amount of current being directly proportional to the difference in temperature of the two junctions.

One of the earliest uses of the thermocouple to measure radiation from long period variables was made by Pettit and Nicolson who employed microscopically thin bismuth-zinc couples in an evacuated tube placed at the Newtonian focus of the 100-inch reflector at Mount Wilson. The window of the tube consisted of rock salt since although glass is transparent to both visible and ultra-violet light, it effectively cuts out much of the infra-red radiation. In the interval 1921 to 1927 they measured the radiometric magnitudes for nineteen long period variables of spectral type Me and two of type Se, the system of radiometric magnitudes being chosen such that the visual and radiometric magnitudes are equal for the bright stars of spectral type A0.

Here we should mention one of the ways in which the thermocouple is superior to the photoelectric cell or the photographic plate in the estimation of magnitudes. Unlike either of these two methods, the thermocouple is equally sensitive to radiations covering a wide spread of wavelengths, in particular to the infra-red region of the spectrum which has no effect at all on the photocell or the silver emulsion of a photographic plate. As a result we may obtain a truer picture of the total radiation emitted by a star by means of the thermocouple.

Pettit and Nicolson's results based on eleven stars of which six were observed at all phases showed a mean radiometric magnitude range of 0.9 magnitude corresponding to a visual range of 5.3 magnitudes. An interesting feature of the radiometric curves (Fig. 27), is that the maxima of the radiometric and light curves do not coincide, the former occurring about a tenth of a cycle later than the visual maximum. This has been found to be a general characteristic of the long period variables.

Further valuable information was obtained by placing a standard water cell in the light path of the variable and comparing the magnitude of the transmitted light with that of the comparison stars. The light transmitted by a water cell includes the visual and the near infra-red region and comparison of the curve plotted from these observations shows that the maximum intensity of light transmitted by the cell occurs about

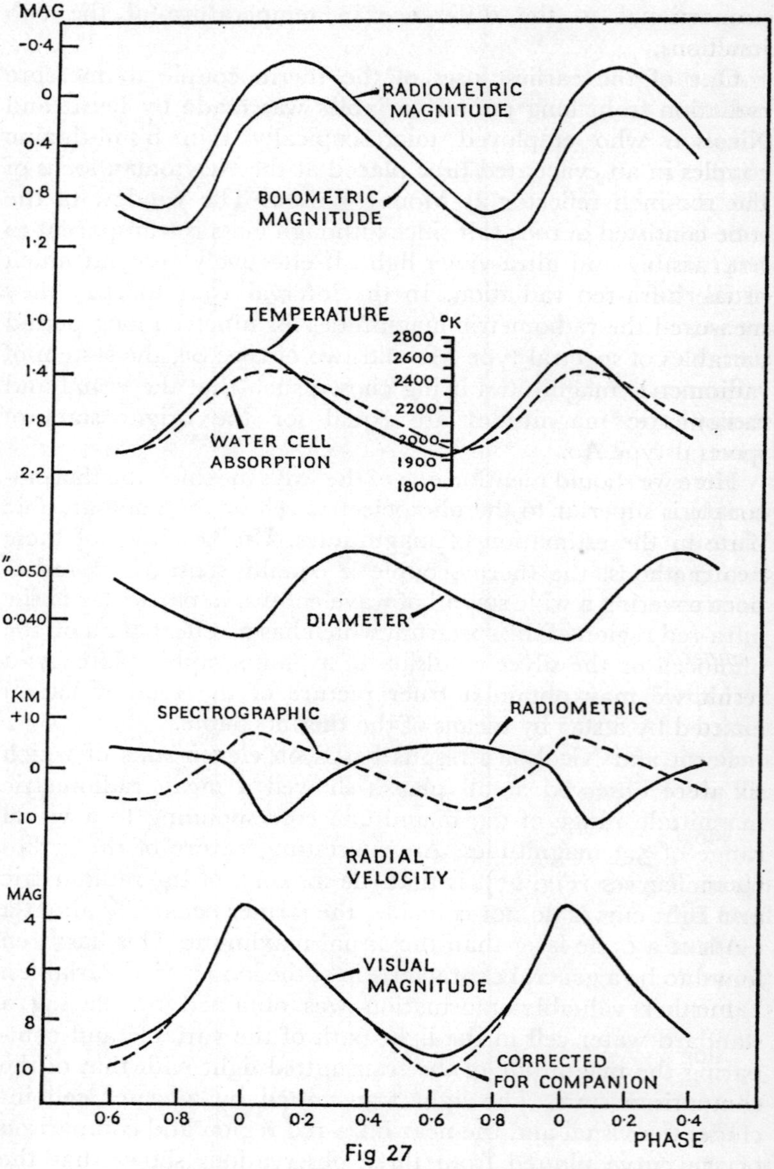

Fig. 27 – Light and radiometric curves of ο Ceti (*Astrophysical Journal by kind permission of the University of Chicago Press*).

0.04 cycle after the visual maximum. This is an important result since it indicates that the maximum of emitted energy from the star occurs later in the cycle when we measure radiation of longer wavelengths and current opinion regards this as due to the lower contribution made at these longer wavelengths by the broad absorption bands of titanium oxide. The maximum of the total energy emitted, taking into account such factors as atmospheric absorption, seems to occur about 0.14 phase or between forty and sixty days after maximum visual brightness, since the absorption bands of titanium oxide are weakest at visual maximum and not at the time of maximum energy emission.

From the foregoing discussion it will be appreciated that the energy concerned in the large amplitudes shown in the light curves of these variables is but a small fraction of that emitted by the star. To a certain extent this is true for all classes of variable stars but it is far more pronounced in the case of the Mira-type variables since, being cool and red objects, the greater part of their energy will lie in the infra-red region of the spectrum.

Temperatures and Diameters from Radiation Measurements

Perhaps of more importance than the amount of light transmitted by the standard water cell is that which is absorbed since the temperature of a star may be found from the water cell absorption by means of the following simple equation,

$$\log T = 3.477 - 0.623 \log WC \qquad (1)$$

where T is the temperature and WC is the magnitude derived from the water cell absorption.

Assuming too that the star radiates as a black body and shows no limb darkening (neither of which is strictly true), the diameter D is given by the relationship,

$$\log D = 1.246 \log WC + 0.0468\ WC - 0.2\ mr - 1.628 \qquad (2)$$

where mr is the radiometric magnitude.

The above equations do not hold for stars earlier in the spectral sequence than gG6 or dK2 since for temperatures in excess of 6,000°K equation (1) does not provide even approximate values of the temperature whereas at 2,000°K the maximum deviation is estimated as only about 10°K.

Using the results obtained from the six long period variables most thoroughly investigated, Pettit and Nicolson found that the mean diameters were 18 per cent larger at minimum than at maximum light, providing a further confirmation for the pulsation theory.

We must now take into consideration the fact that these stars do not strictly follow the black body radiation law nor is there any evidence to lead us to suppose that limb darkening is absent. As mentioned in Chapter 1, limb darkening is pronounced in the case of the sun and there is strong observational evidence of it in certain eclipsing variables. It is only reasonable therefore to assume that the same is true of the long period variables but as yet there is no way by which the magnitude of this effect may be computed. If we go on the assumption that these stars behave similarly to the sun, then the calculated temperatures must be increased by about four per cent to give us the effective temperature at the centre of the stellar disc. The calculated diameters however will be only slightly in error because of the manner in which they have been derived.

Even the results obtained so far do not exhaust all of the information we may obtain from a study of the various kinds of radiation curves. Like the extrinsic variables discussed earlier in the book, certain of the long period variables are known to have close companions although only rarely do eclipses occur. This fact was discovered spectroscopically at the beginning of the century. Now it is known that the heat index of a star (the difference between the visual and radiometric magnitudes) may be found from the water cell absorption and from this the true visual magnitude at minimum phase. In Fig. 27 the dotted lines in the visual light curve show the variation in magnitude of the variable corrected in this way for the unseen companion.

As an example, we may take the case of *o* Ceti. The heat index at maximum light is 3.67 magnitudes and is not affected by the presence of the companion but is 1.1 magnitude greater than normal for its water cell absorption. The heat index at minimum is 8.0 magnitude derived from a water cell magnitude of 2.04. This increases to 9.1 magnitude when we take into account the correction of 1.1 magnitude found at maximum and since the radiometric magnitude at minimum is found to be 0.86 this gives a visual magnitude at minimum of 9.96 compared with the observed visual magnitude of 9.00 which is, of course, the combined magnitude for the two stars. From this we obtain a figure of 9.6 for the brightness of the unseen companion star which agrees very closely with a value of 9.8 magnitude obtained by independent methods.

Spectra of Long Period Variables

From the very early days of stellar spectroscopy, the long period variables have provided astronomers with a rich source of spectroscopic data. It is found that during their light variation, considerable changes take place in their spectra giving us a detailed insight into the physical changes of the star. Since they are all cool, red giant or supergiant stars we would naturally expect them to be of spectral class K or later; in fact very few having K type spectra even at maximum brightness are known.

By far the greater proportion of these stars have M type spectra showing bright emission lines of hydrogen around maximum and the presence of these bright lines was originally thought to indicate a fundamental difference between these stars and the semi-regular and irregular variables although later work has suggested that such a difference, if it exists at all, is very slight.

Secchi in 1866 included the long period variables in class III of his spectral classification; a class which he later subdivided into classes III and IV, the former having the fluting of the broad absorption bands shaded towards the red and the latter with flutings shaded towards the violet. With the introduction of the Harvard sequence of stellar spectra, Secchi's class III was classified as Types M and S stars, and his class IV as the R and N stars.

The M stars are characterised by the presence in their spectra of broad absorption bands of titanium oxide since the temperature range of these variables is insufficient to bring about the dissociation of these relatively stable molecules. For the same reason, the normal lines due to the neutral metals are also very numerous, those of calcium and iron being particularly intense. The absorption lines due to ionised metals are weaker than in the preceding class K due to the lower surface temperature.

The lines of hydrogen are almost invisible except when they appear in emission around the time of maximum brightness. These stars too are found to have a low carbon to oxygen ratio and unlike the N and R variables, most of the carbon in their atmospheres appears to have been used up in the formation of the very stable but spectroscopically unobservable carbon monoxide molecule, the remaining oxygen then being available for the production of titanium oxide. As will be seen later in the discussion of the semi-regular variables, the fact that titanium oxide absorbs strongly at the red end of the spectrum means that as these stars fade and more and more of the red end is absorbed, the colour of the star does not become appreciably redder towards minimum, the colour index remaining almost constant throughout the entire light cycle.

The spectral type S was added in 1922 to indicate that although these variables closely resemble the M stars, zirconium oxide bands are here stronger than those due to titanium oxide. One must not imagine however that there is a sharp distinction between them. Stars are known with various relative strengths of both absorptions and it is quite possible that there is even a continuous sequence of stars having spectra ranging from M to S. The two best known Mira type variables having S type spectra are R Andromedae and R Geminorum, both of which have been studied in detail for more than a century.

In addition to zirconium and titanium oxide bands, absorptions due to other fairly stable molecules are present including yttrium oxide, lanthanum oxide and silicon hydride. Other absorption lines are enhanced as compared with those in the K or M types stars.

The majority of these variables are quite distant objects but from the proper motions of seventeen of them Keenan has estimated a mean absolute magnitude of −1.0 and accordingly they are normal giants. At the present time no dwarf or supergiant stars of type S are known, perhaps because of their comparative rarity. As a group they possess low velocities and have a distribution in the galaxy very similar to the N and R type stars. Although few are known it may well be that their total number is fairly large, the main difficulty being in recognising them.

One of the most remarkable spectral peculiarities of these stars is undoubtedly the discovery by Merrill in 1951 of the absorption lines of technicium, a very unstable radioactive element first identified in the laboratory as recently as 1937. No stable isotopes of this element are known and the most stable has a half-life of only 200,000 years. The absorption lines due to technicium in the neutral state have been positively identified in the spectra of both R Andromedae and R Geminorum as well as in several other S type variables. No fully acceptable explanation of the existence of this extremely rare element in these stars has as yet been given.

The M and S stars are so very similar that an explanation based on purely physical conditions prevailing in these variables cannot possibly explain the side range of relative strengths of titanium and zirconium oxide bands in their spectra. The dissociation temperatures of these two molecules are slightly different it is true and although some of the changes in line and band intensities may be due to physical changes, astronomers are now coming round to the opinion that the major cause is to be found in the relative abundances of the various chemical elements.

Within the past ten years our knowledge of the processes operative within stellar interiors has provided us with a reasonably clear picture of the way in which the heavier elements are built up from those lighter than iron by means of neutron capture and beta decay reactions. Some of these will be explained in greater detail in the chapter on supernovae. One further point about the bright emission lines of hydrogen found at maximum light must be made here. Many of them are found to have absorption components superimposed upon them showing that the emission lines originate fairly deep inside the stellar atmosphere. It seems likely that in these stars there is a hydrogen convection zone beneath the photosphere which becomes increasingly visible as the variable approaches maximum light.

Let us now examine the two other types of spectra found among the long period variables namely types N and R. The broad absorption bands in these spectra are due to carbon compounds and unlike those of the two previous classes absorb strongly in the violet. As a consequence, these stars become progressively redder as they approach minimum due to the increasing intensity of the bands in the blue and violet allowing the red end of the spectrum to dominate.

Such carbon stars exhibit strong bands due to C_2, CN and CH molecules and in certain cases bands of NH are also found. Type N stars are characterised too by a group of bands previously discovered in the heads of comets and recently shown to be due to the linear triatomic C_3 molecule.

A surprising fact which has emerged from the study of these variables is that the relative abundance of the two main isotopes of carbon is far different in these stars to that found on Earth or even in the sun. Search for the rare carbon-13 isotope in the sun has so far proved unsuccessful while here on Earth, the carbon which makes up much of living matter, including the cells of animals and human beings consists of at least 90 per cent of the carbon-12 isotope, only the remaining 10 per cent being the rarer carbon-13. In these stars the ratio is nearer 50:50.

No dwarf stars of types N or R are known, all being either giants or in a small number of cases, supergiants. From radial velocities and proper motions, mean absolute magnitudes have been determined, these being -1.8 and -0.5 for type N and R respectively.

As opposed to the M and S type variables there is undoubtedly a large amount of carbon present in proportion to oxygen and it is the latter which is all used up to form carbon monoxide, leaving an excess of carbon which is then able to form the various carbon and hydrocarbon molecules responsible for the absorption spectrum. Not all of the N and R stars have the same carbon to oxygen ratio and it is difficult to understand why there should be an excess of carbon at all unless we assume that it is produced in the centre of the star as a product of helium burning. Certainly the necessary nuclear reactions for the formation of carbon from helium are well known but we do not at the moment know the kind of mechanism which would explain how the carbon, once it is formed, is transported to the surface layers of the star. From the observed abundance of the carbon-13 isotope it appears inescapable that the Bethe carbon-nitrogen cycle must be operative in the interior.

Two final points must be made here concerning the bright emission lines in the spectra of these stars. So far as can be seen from the data at present available, the hydrogen emission lines behave in a very similar manner to those in the RR Lyrae and Cepheid variables, appearing on the ascending branch of the light curve and fading some time after light maximum. Almost invariably Hδ is seen first, followed by the lower Balmer lines with Hγ becoming eventually the strongest member of the series. One star, X Monocerotis, is peculiar in that the emission lines are present throughout the whole of the light cycle although varying in intensity. At the present time we do not know why this should be so, either the emission cycles overlap in this particular variable or it may be that the physical conditions producing emission have somehow become stabilised to a certain extent in the atmosphere of this star.

Lastly we again have the linear logarithmic relationship between period and the time during which the calcium emission appears which we have already seen holds for the Population I Cepheids.

Radial Velocities of Long Period Variables

Since we are here dealing with giant stars (in a few cases even supergiants), with very deep, tenuous atmospheres, it is scarcely surprising that the radial velocity curves found for the long period variables do not exhibit the same regular nature shown by those of the RR Lyrae and Cepheid variables. However some generalities have been deduced from a study of more than one hundred of these stars, mainly of spectral types M and S, some being rather vague as might be anticipated from the wide range of period, amplitudes and spectra concerned.

(1) In all of the stars which have been investigated, the radial velocities found from the absorption lines are systematically higher than the corresponding ones derived from the emission

lines. Since the absorption lines originate in the deeper levels of the stellar atmosphere where the velocity of a pulsation front or a shock wave will be larger (before the energy has been dissipated), this is a result we would normally expect. In these stars, too, there is a general correlation between the period and the velocity difference of the absorption and emission radial velocity curves.

(2) The radial velocity obtained from the hydrogen emission lines is invariably lower than that found from the emission lines of the metals and there is some evidence, admittedly flimsy, to show that the velocities of the individual Balmer lines of hydrogen differ among themselves. It is possibly too early yet to say whether these latter small differences are real or not.

Apart from the above correlations, the radial velocity curves for the individual long period variables show fairly pronounced differences which do not appear to depend on either period or spectral type. Many show an almost constant radial velocity from the absorption lines (which is almost certainly that of the centre of gravity of the star itself) for a large proportion of the period, covering the latter half of the rise to maximum and lasting for about a tenth of the period after maximum when there is a sudden decrease corresponding to the ejection of mass. As we shall now see this is consistent with the formation of a shockwave in the stellar atmosphere.

Shock Waves in Long Period Variables

The theory of shock-like disturbances in the very extended atmospheres of these stars has been developed over the past two decades and explains many of the observed spectral phenomena together with certain features of the light curves of these variables. From the changes in the radial velocity curves, it is evident that the velocities associated with a shock wave front are quite small being of the order of only 10 to 15 kilometres per second, only about a tenth of the velocities found in the β Canis Majoris stars for example.

It must be borne in mind however that the atmospheres of these giant stars are both cool (surface temperatures between 1,800 and 2,600°K) and exceedingly tenuous and under these conditions, assuming the composition to be virtually all atomic hydrogen, the speed of sound will be around four kilometres per second, about a third of that of the observed disturbance.

A greatly simplified picture of the sequence of events is as follows. A shock wave is generated below the reversing layer within a hydrogen convection zone and moves outward, passing through the reversing layer, imparting an outward motion to these gases. Beyond the reversing layer, the density decreases quite rapidly and accordingly the velocity of the wave front increases even though the energy of the wave is now rapidly being dissipated. Owing to the rarity of the matter in these giant stars, the gravity is very low (at the surface it is about the same as that which we experience here on Earth) and as a result the wave front can travel outward for a large distance before all of the energy is lost and it comes to rest.

The gases through which the shock wave is passing are, of course, neutral but since one result of its passage will be to raise the temperature, this gas will naturally become ionised although recombination of these ionised atoms will take place quite rapidly. Judging from a comparison of light and radial velocity curves, it can be assumed that maximum brightness is coexistent with maximum strength of the shock wave.

Masses of Long Period Variables

On the Russell-Hertzsprung diagram, the long period variables occupy a region of instability to the right of the main sequence and it is therefore possible to estimate their masses from evolutionary considerations. Long period variables have been found in fairly large numbers in at least one of the globular clusters (47 Tucanae) and for these it is possible to use the Russell-Hertzsprung diagram compiled from the stars in this cluster to estimate both the masses and the age of these particular variables. In general these long period variables, which are high velocity stars with short periods, usually less than 250 days, and Population II stars have masses similar to that of the sun and ages around 15,000,000,000 years.

The longer period variables are generally Population I objects and possess lower velocities than the former; they have also larger masses between 2.0 and 5.0 solar masses and in agreement with their Population type are younger stars with ages estimated at around 600,000,000 years. A star will inevitably lose some of its mass during its initial stages of evolution but it is not easy to assess how much although it is doubtful if it will be in excess of 0.5 solar mass. It must not be imagined however that the long period variables all fall into these two distinct classes; these must only be looked upon as the extremes. The vast majority of these stars have properties which are intermediate between these two types.

There are, in addition, some uncertainties about the masses as derived in this manner. We have already seen that only in the case of binary systems can we obtain accurate estimates of mass and although some binaries are known in which one component is a long period variable; e.g. Mira, R. Aquarii and X Ophiuchi, the elements of their orbits are not at present known with the precision of the Algol variables. The masses obtained from a study of the orbits of these binaries are however in fairly good agreement with those derived from evolutionary considerations but provide us with no significant improvement in accuracy.

11. Semi-regular Variables

The first variable of this class to be discovered was the third magnitude star α Herculis which was found by Sir William Herschel in 1796 to vary between 3.0 and 4.0 magnitude, the variation in light being seemingly of a totally irregular kind. John Herschel who continued the observation of several classes of variable stars called attention to many variables which apparently had no regular period, not even of the type exhibited by the long period variables typified by Mira although he was well aware of the many irregularities in this class of star which he attributed to some semi-periodic veiling effect of interposing dark nebulous matter.

Throughout most of the eighteenth and nineteenth centuries, the observers of variable stars were unfortunately working with very meagre data and logically or otherwise, they became convinced that in every type of variable, a period of some kind existed in each case although they considered that it might be concealed beneath a variety of minor fluctuations and hence not easy to elucidate. In addition, since only a relatively small number of variables were known at that time there was the natural tendency to examine each star individually rather than to attempt to form them into discrete classes and view their properties as a whole and it was not until the end of the nineteenth century when a large mass of data had been painstakingly accumulated and a far larger number of variables had been assigned to certain definite classes, that it emerged there still existed some stars for which very little, or no, evidence of periodicity in their light curves could be uncovered. The stars for which, even today, no evidence of periodicity has been found are the true irregular variables which will be discussed in a later chapter. Those for which a certain amount of periodicity can be found, albeit at times of a somewhat transitory nature, are classed as the semi-regular variables.

It is, of course, very difficult to define a period for these stars since by their very nature they possess an irregularity even more marked than that of the long period variables just discussed where we have seen that the observed period can differ from the mean by more than a month in many cases and individual maxima by more than three magnitudes.

Like the long period variables, these stars are mainly red ones of late spectral type and all are giants or supergiants. Their amplitudes are smaller than those of the Mira variables rarely exceeding two magnitudes and as we might expect their light curves are very diverse. Any subdivision of this large and heterogenous class must be somewhat arbitrary but four sub-classes are now generally recognised. These are as follows:

SRa: Giant stars having late type spectra. The light curves of these variables are often very similar to those of the long period variables, especially those of the latter whose behaviour is even more irregular than that of Mira. Variables belonging to this sub-class such as S Aquilae and R Ursae Minoris were originally classed as long period variables illustrating the great similarity between the two. Quite often these stars may be differentiated from the true long period variables only by their smaller light range.

SRb: These stars exhibit a generally recognisable periodicity but at times this is replaced by slow, more irregular variations and there may even be periods when the star is of almost constant brightness. U Boötis, a star which has been constantly followed by the Variable Star Section of the British Astronomical Association, is typical of this group.

SRc: A fairly large group of variables, all supergiants of late spectral classes, typified by α Herculis and α Orionis (Betelgeuse). The light changes are normally of the order of a single magnitude.

SRd: This is a small group of giant and supergiant variables having earlier spectra than those of the preceding class, usually between types F and K. Many of them have light curves very reminiscent of the RV Tauri variables which will be described in the following chapter with alternate deep and shallow minima.

Light Variations of Semi-regular Variables

In spite of the difficulties outlined above it is possible to find a certain degree of regularity in the light curves of these stars but this cannot be strictly described as the period of the star since this would imply that there is a reasonably constant recurrence of the phase in question. In the case of the semi-regular variables, except for certain stars of groups SRa and SRb, there is not even an approximately constant interval between successive maxima (or minima) and often there are the long periods of completely irregular behaviour including times when the star may remain at virtually constant brightness, which complicate matters even further.

Nevertheless, it is possible to find a certain phase interval for most of these variables which recurs more often than others, giving a frequency maximum or cycle for each star. Thus instead of a mean period as we have for the Mira type variables, the semi-regular class of stars may be characterised by a cycle indicating that there does exist a form of periodicity but one which is combined and obscured by irregular variations in brightness.

It is particularly unfortunate that initially little interest was taken by astronomers in the semi-regular variables. The reason for this lack of enthusiasm is not hard to find. The early observers attempted to find periods for such stars and when this either failed or grossly conflicting results were obtained they were simply classed as 'irregular' and little further serious work was undertaken. Had they been followed as assiduously as the long period variables we would now have had much more extended light curves providing important information about these stars, information which is in the process of being accumulated at the present time.

Since the end of the last century an increased interest has been taken in these variables with the result that detailed data concerning their light variation has been gained bringing to light many interesting features. In many instances, the investigations have shown that a very complicated periodicity is present consisting of two or more periods superimposed upon each other.

If we take the shortest cycle for these stars to be the primary one, i.e. the cycle which is intrinsically related to a possible pulsation of the star itself, then a frequency curve may be constructed for these stars as shown in Fig. 28 which is based upon observations for 316 stars of this class.

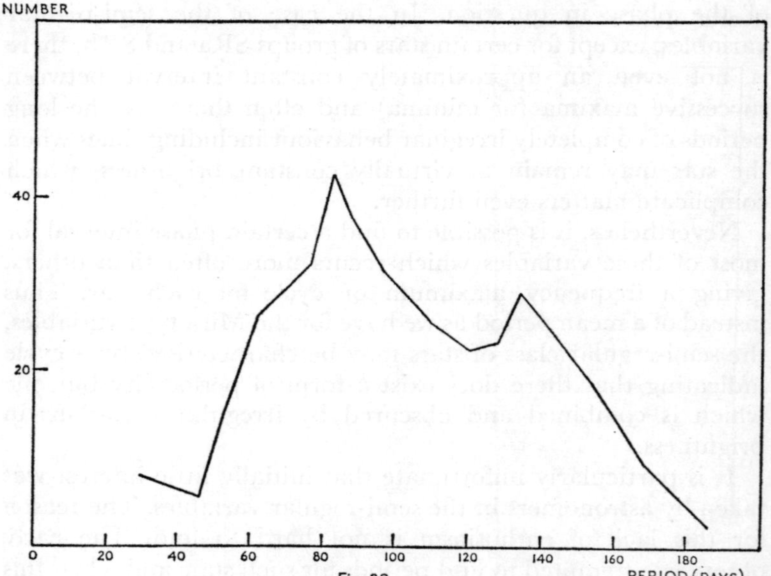

Fig. 28 – The frequency-distribution curve for semi-regular variables.

It will be seen from the above figure that there are two maxima in the distribution curve occurring around 85 days and 135 days but it must be emphasised that our knowledge of these stars is still very far from complete and it is not possible at the moment to say with certainty that these apparently favoured periods indicate a real physical difference among these variables.

Comparing the light curves of the semi-regular variables with those of the Cepheids, or even the long period variables, we find one basic difference, namely that whereas the latter show quite a strong family resemblance from one star to another, this is not so for the semi-regular stars. Here there are large variations in the curves obtained for individual stars, some being far more regular than others.

The nature of the curves seems to suggest the presence of at least one long wave superimposed upon the basic cycle but in every case long and continuous observation is necessary to reveal the presence of this slow wave since it is normally one of several years duration. UZ Persei which is one of these variables

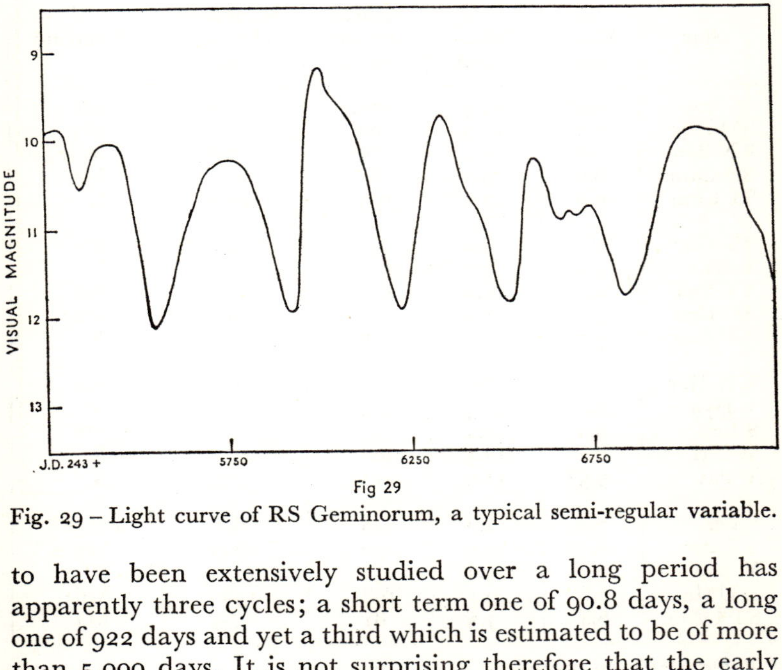

Fig. 29 – Light curve of RS Geminorum, a typical semi-regular variable.

to have been extensively studied over a long period has apparently three cycles; a short term one of 90.8 days, a long one of 922 days and yet a third which is estimated to be of more than 5,000 days. It is not surprising therefore that the early investigators failed to find any regularity at all in its light behaviour; the number of observations made being far too few and too sparsely scattered to allow of any detailed analysis of the light curve.

In general therefore the apparent irregularities in the light curves of these stars may be explained as a result of a very complex variation brought about by the interference of two or more oscillations, each of which may possibly be more or less regular and varying independently of the others. To separate these individual oscillations, the method of harmonic analysis must be employed and this requires long series of uninterrupted observations which have as yet been made for only a small number of these stars. For the large majority the presence of

Table 9
SEMI-REGULAR VARIABLES

Star	Maximum	Minimum	Primary Period (days)	Secondary Period (days)	Spectrum
SS And	10.0*	11.4	152.5	650	M6
BM Aqr	11.0*	11.9	60	550	M4
RX Boo	6.9	9.1	78	500	M8e
RS Cam	8.0	9.6	85	960	M5
ST Cam	9.0	12.1	195	2100	N
RT Cnc	7.3	8.6	94.5	540	M5
Y CVn	5.2	6.6	158	2100	N
VY Cas	9.0	10.2	100	620	M6
AD Cen	9.8*	11.4	62	2800	M1e
RV Cyg	7.1	9.3	75	470	N
UW Her	7.5	8.6	81	195	M5
V Hya	6.0	12.5	532	6500	N
S Lep	6.0	7.4	95	835	M6
W Ori	8.2*	12.4	200	2300	N
V Pav	9.3*	11.2	225.5	3735	N
S Per	7.2	12.2	810	916	cM3e
TT Per	9.3*	10.5	90	843	M5
Y Tau	6.8	9.2	241	1750	N
Z UMa	6.6	9.1	198	1560	M6e
ST UMa	6.4	7.5	81	583	M4

* Photographic magnitude

such a long, secondary variation, observed as a second maximum, is characteristic of their light curves. This maximum progresses along the light curve relative to the primary cycle with the result that the amplitude varies continuously from a minimum to a maximum value. The light curve of RS Geminorum, shown in Fig. 29, is typical of such behaviour.

In addition to the four sub-classes mentioned earlier, two general groups of semi-regular variables may be distinguished:

(a) Those in which there is a variation in amplitude which may or may not be accompanied by a corresponding change in the mean brightness such as TZ Draconis and,

(b) variables which exhibit short-cycle variations on a long period wave, e.g. VY Cassiopeiae and RR Coronae Borealis.

Table 9 lists some semi-regular variables for which values of the two cycles have been determined although it must be pointed out that the figures given for the long-term cycle are often highly uncertain because of lack of information and the small number of observations on which they are based.

It is of interest to note that those variables having N type spectra do in general have the longer primary and secondary cycles.

Spectral Characteristics

We must here consider the spectra of the semi-regular variables and in particular how they differ from those of the long period variables. We have already seen in the previous chapter that the Mira type stars belong almost exclusively to the spectral classes M, N, R and S and in general the same may be said for the semi-regulars. A small number are known of earlier spectral type, the SRd variables, but by far the larger majority are M or N type stars.

At first sight the main spectral difference between the two classes would appear to be the appearance of bright emission lines of hydrogen in the spectra of the Mira variables. Indeed, at one time, it was considered that the presence of these lines indicated some fundamental difference in the composition and physical properties of these stars which, in the case of the long period variables, might be responsible for the greater stability of the pulsations taking place in the outer atmospheric layers.

This suggestion naturally leads us to consider whether emission lines are absent from the spectra of the semi-regular variables because of some basic change in their constitution or whether they do actually occur but owing to difficulties of observation they can only rarely be identified. From an examination of Table 9 it would seem that the number of semi-regular stars having emission spectra is very small and for a long time such lines were not observed even though a great many of these stars are known and in their general behaviour they appear to be quite closely related to the long period variables.

Let us however look at the position a little more closely. It will be recalled that in the spectra of the Mira type variables, the bright hydrogen emission lines are not present throughout the whole of the light cycle. They appear about halfway from minimum to maximum brightness and disappear before the following minimum. Generally speaking therefore, they can be identified for only about a third of the whole period even in the most favourable cases. Because of the quite large amplitudes and long periods of these variables however we would expect the probability of finding emission lines in their spectra to be extremely high since they will be visible for several months even in the case of those stars with the shortest periods.

The semi-regulars on the other hand have much smaller amplitudes and shorter primary cycles and hence the chance of observing any emission lines which may be present is appreciably smaller. Since the visual estimate of the brightness of a variable can be made in a matter of minutes much of the earlier work on these stars consisted of the production of their light curves, a task well suited to amateur observers and consequently we know quite a lot about their light variation.

To obtain a satisfactory spectrum however may require an exposure of several hours and the amount of work involved in following the spectral changes of these stars is correspondingly greater. As a result it was not until long and detailed spectral studies of these variables had been carried out that emission lines were observed.

μ Cephei is one of the brightest of the semi-regular stars and has been so extensively observed by many astronomers that it will be described more fully in a later chapter. At present it will be sufficient to say that at the very bright maximum in 1930, it was discovered that both Hγ and Hδ appeared as emission lines with a displacement relative to the absorption lines towards the red end of the spectrum amounting to a velocity of recession of 3.5 kilometres per second. One strange feature of these bright lines was that this displacement appeared to vary with the changes in magnitude.

Another peculiar semi-regular variable is SX Herculis which is a member of the small group of SRd variables having a K type spectrum at maximum and showing light variations which place it intermediate between the long period variables and the semi-regulars. Here too, bright lines appear shortly after minimum light, attaining their maximum intensity a little before maximum brightness and then decreasing in strength quite rapidly, disappearing altogether about forty days before the following minimum.

A third variable worthy of note here is S Vulpeculae, also of the sub-group SRd which has characteristics intermediate between the Cepheids and the present class. The light curve shows quite definite Cepheid-like variations and the mean cycle of 67.77 days and spectral type cK7v can be made to fit into the period-luminosity curve. Like the Cepheids, the visual range of about 0.5 magnitude is less than the photographic range of 1.2 magnitudes. However there is no simple correlation between the light and radial velocity curves as is invariably shown by the Cepheids and the individual cycles deviate considerably from the mean.

From what has just been said, we can see that the early belief that the presence or absence of hydrogen emission lines in the spectrum is sufficient to differentiate between the long period and the semi-regular variables was based upon insufficient observational data. The generally accepted view now is that both classes of star are spectroscopically very similar although there is some evidence that emission lines in the spectra of the latter class do not follow exactly the same pattern throughout the light cycle as those of the long period variables. On the other hand, there is still the possibility that some of the semi-regular variables do not exhibit emission lines in their spectra at any phase of the light curve but we must await the accumulation of more numerous and comprehensive observations taken over a long period of time before we can reach any definite conclusions on this point.

Although the presence of emission lines in the semi-regular stars has now been demonstrated, it may nevertheless be asked whether there is not some other difference in their spectra, for example the distribution of the stars among the different spectral types. A survey of several hundred variables of each class has shown that the carbon stars of type N are more numerous among the semi-regular variables than among the Mira type stars. Almost half of the known semi-regulars are of this spectral class, the remainder being of type M; very few indeed of types S and R are known.

Circumstellar Envelopes Around Semi-Regular Variables

Since most, if not all, of the semi-regular stars are giants or supergiants, similar to the RV Tauri stars in this respect, we must now consider whether there is any evidence from their spectra for the presence of huge circumstellar envelopes surrounding these variables. The brightest of this class is the beautiful red star Betelgeuse in the shoulder of Orion, varying between 0.4 to 1.3 magnitude in a mean cycle of about 2,070 days. Interferometric measurements made with the 100-inch reflector at Mount Wilson have provided a mean diameter of 300,000,000 miles for this supergiant star, large enough to include the whole orbit of the Earth, and it has been conclusively established that variations in the diameter occur with the changes in brightness. Similar changes have been found also for the first magnitude star Antares (α Scorpii).

High dispersion spectrograms taken in 1935 by Adams and MacCormack showed that the strong resonance lines of Betelgeuse possess absorption cores which are displaced towards the violet, such features being subsequently discovered in many other M type supergiants. Spitzer has suggested that in the outer layers of the grossly distended atmospheres of these vast, extremely rarified bodies huge fountains of gas, possibly similar to the prominences seen on the sun although on a much larger scale, erupt from the surface and become ionised before falling back into the body of the star. Theoretical considerations lead us to believe that during the evolution of these massive stars, ejection of matter will occur, almost certainly during the red supergiant phase of such a star's career.

Spitzer's model has been greatly modified in recent years. It is now considered, as first put forward by Deutsch, that matter is being continuously ejected from the surface of these stars, forming an envelope of enormous extent around them. Some idea of the size of these circumstellar envelopes may be gained from a study of α Herculis, the first semi-regular star to be discovered. Fortunately this particular variable happens to be a binary system comprising the M type supergiant and a G type yellow star. The presence of this companion star enables us to obtain additional information concerning the size of the circumstellar envelope around the supergiant since, if the circumstellar lines in the G class spectrum are approximately as strong as those in that of the M star, we may conclude that the gaseous envelope extends out at least as far as the yellow component.

Unfortunately, the spectral lines are not as clear in this system as they are for Betelgeuse and the results are not yet conclusive on this point. The evidence we have at present does suggest that the G star is indeed embedded within this huge envelope which surrounds the supergiant star. If this is so, then the dimensions of this envelope must be tremendous indeed. Deutsch's original finding that in the α Herculis system the matter within the envelope is agglomerated into relatively dense lumps has been recently re-examined. Discontinuities such as he has suggested are not unknown but it would appear from the more recent determinations that if they do exist, they are not as pronounced as was at first thought.

The concordant results derived theoretically and observationally leave no doubt about the existence of these vastly huge gaseous envelopes surrounding these variables; what is not yet known is the actual mechanism responsible for the continuous ejection of matter from the surface of the star. Quite probably there are powerful turbulences within these tenuous atmospheres, possibly produced by shock waves similar to those postulated for the long period variables which serve to keep the outermost layers in a continual state of motion away from the main body of the star with doubtless some of the material escaping away into space.

Surface Temperature

It is obvious at once from the fact that almost all of the variables of this class are red stars (only a small number being yellow or orange) that their surface temperatures must be fairly low, certainly 3,000°K or less. One method of deriving a reasonably accurate estimate of their surface temperatures has already been mentioned, namely by determining the proportions of the various colours in the spectrum using a very sensitive photometer and comparing the results with the colour distribution in the spectrum of a standard lamp whose temperature is accurately known by laboratory experiment. Another way is to measure the bolometric magnitude at various phases – that is the total amount of radiation emitted by the star, taking into account not only the visible light but also the infra-red and ultra-violet radiations not visible to the eye and allowing for those parts of the electromagnetic spectrum which are absorbed by the gases in the terrestrial atmosphere.

On the basis of the bolometric data thus obtained for several semi-regular variables it is shown that the surface temperature varies from a maximum of 2,400°K to a minimum of 1,800°K for the M type stars. Those having N type spectra appear to have a slightly lower maximum temperature around 2,250°K. There are, of course, several difficulties in the precise measurement of these low temperatures and the above data can only be regarded as giving average values based on only a small percentage of the total number of semi-regular variables known.

The surprising feature about these stars is the very small temperature range compared with their visual amplitude, a characteristic which has already been mentioned in connection with the long period variables where the difference is even more striking. Here too it is found that the bolometric magnitude reaches a maximum about a tenth of a period after maximum visual brightness whereas the maximum temperature corresponds almost exactly with the maximum in the light curve.

The most likely cause of the temperature changes in these stars is that of pulsation, possibly coupled with shock wave phenomena. The former has been measured directly from interferometric measures of the diameter of Betelgeuse and Antares. As we have seen, Betelgeuse has a mean diameter about 330 times that of the sun and at a distance of 296 light years is sufficiently close for its diameter to be measured directly. Pease, in 1926, was able to demonstrate that the diameter is not constant but varies between 300 and 400 times that of the sun, the diameter being around maximum when the star is faintest.

The SRd Variables

Before going on to consider the galactic distribution of these variables, it is important here to briefly discuss the yellow semi-regular variables of the sub-class SRd since their light curves are so similar to those of the RV Tauri variables that it is often difficult to differentiate between them. Here we must anticipate a little in order to put the present picture into its proper perspective.

The RV Tauri stars form a very heterogeneous class of highly luminous stars having light curves possessing alternate deep and shallow minima and periods between 45 and 170 days. Their spectral types at maximum are always types F, G or K although around a deep minimum the broad absorption bands of titanium oxide, equivalent to an M type spectrum, may be found.

The yellow SRd variables also satisfy, to a large extent, the three criteria mentioned above and until about ten years ago, a great deal of confusion existed about these two groups, mainly because of inadequate three-colour photometry and spectroscopic examination of these variables.

The first clue to a real difference between the classes came from comprehensive spectroscopic surveys when the majority of the SRd stars were found to show quite strong and persistent emission lines of hydrogen, features which are so far known in only two RV Tauri variables and even here the hydrogen emission is relatively weak. The absorption spectra are also unlike, those of the SRd stars showing no absorption lines of hydrogen and weak cyanogen bands.

The radial velocities of the SRd variables are also, on average, systematically higher than those of the RV Tauri stars, placing them among the Population II stars, their kinematic properties being quite definitely those of the galactic halo population.

Galactic Distribution

The important question of the galactic distribution of the many classes of variable stars has received much attention of late. For example, it is well known that the novae tend to concentrate along the plane of the galaxy, very few having been discovered in the region of the galactic poles. The same may be said of the Cepheids which show a remarkable tendency to cluster in certain parts of the sky, e.g. Perseus, Orion, Cygnus, Scutum and Carina.

By contrast, the semi-regular variables are found in all galactic latitudes, the number being somewhat larger for the southern than the northern hemisphere. A statistical survey however indicates that, like the Cepheids, there are certain parts of the sky in which they seem to cluster although not on the same scale as the Cepheids. It is fortunate though that many of the semi-regular stars appear in the same regions as the Cepheids since this at once provides us with a means of determining their absolute magnitudes. As we know, we can readily find the absolute magnitude of a Cepheid once we know its period and if we make the assumption that the semi-regular variables are at essentially the same distance we can obtain a figure for their absolute magnitudes with a fair degree of precision. That this assumption is not far from the truth will be seen later.

It will be as well here to mention some of the difficulties in the use of this particular method. Firstly, even in the case of the densest clusters where relatively large numbers of both classes of variable star are found together, the evidence that they are spatially close to each other is only circumstantial. There remains always the possibility that this apparent proximity is due only to the fact that they lie close together in our line of sight. Secondly, it is not easy to define either the maximum or minimum for a semi-regular variable owing to the marked irregularities in the light curve; all we can do is take the extreme magnitudes insofar as they have been observed.

However, provided these limitations to the method are recognised, certain results may be obtained which indicate an average absolute magnitude of -3.5 for these stars.

Are there any other methods we may use to get around this problem? One which immediately suggests itself is to use the globular clusters whose distances are already known from the RR Lyrae stars found therein in large numbers. Since these clusters are extremely compact masses of stars, they would appear ideally suited for this purpose. Unfortunately very few semi-regular variables have been discovered in these clusters and inevitably any statistical analysis will be based on only a small number of stars, too small to be really meaningful.

Two possible candidates have however been found – one in NGC 362, a globular cluster which is situated on the edge of the Small Magellanic Cloud and the other in NGC 6121. Both stars are found to have absolute photographic magnitudes of about −2.0 which, considering the difficulties inherent in both of these methods of determination, is in very good agreement with the values found for the semi-regular variables in our own galaxy.

It is clear, however, even from the limited results we have, that these stars are among the brightest of all variables in the galaxy although at the moment there is not sufficient data available for a period-luminosity curve to be constructed as has been done for the RR Lyrae and Cepheid variables. Considering the complex nature of their light curves, this will not be an easy task, if it is possible at all.

Since we find these stars distributed throughout all galactic latitudes we would expect it to be possible to divide them into two groups, those belonging to the galactic disc (Population I) and those found in the galactic halo (Population II). Of the former, both α Herculis and α Orionis have been extensively observed for more than a century while of the latter, two of the brightest, TY Virginis and SX Herculis, have been studied in detail.

Only a small number of variables like TY Virginis have so far been discovered. Their light curves are erratic in the extreme and for this reason they have often been classed among the totally irregular variables. Like them, they have small ranges of light variation and it is a matter of great difficulty to allocate even a mean cycle for their light changes. Attempts have been made to derive radial velocity curves for these stars but without much success, any deviations from the mean being very small and difficult to dissociate from errors of observation.

SX Herculis, on the other hand, has quite a large visual amplitude and the radial velocity curve is well defined, varying between +10 and +20 kilometres per second which is a range very similar to that of the Cepheids. Maximum velocity occurs a little before maximum light and shows quite clearly that the diameter of the star is at its greatest when it is around minimum brightness.

Measurements of the proper motions of these stars have shown too that they possess very high velocities with respect to the sun, a result we would expect for stars of the galactic halo population. As to their absolute magnitudes, these have been calculated from the emission lines of singly ionised calcium in their spectra and shown to be quite like those of the red giant stars previously found in many of the globular clusters.

Semi-Regular Variables in Other Galaxies

So far we have dealt only with the semi-regular variables within our own galaxy and the attendant globular clusters. What of similar stars in other galactic systems? Do their properties compare with those already discussed or are there any fundamental differences?

We have already seen that large numbers of Cepheids are present in the extragalactic nebulae, these providing us with an important method of determining their distances, at least out to about four million parsecs (thirteen million light years). RR Lyrae variables too abound in the Magellanic Clouds and have also been detected in some of the nearer spiral nebulae.

It would appear highly likely therefore that we can also detect semi-regular variables in these external galaxies since they are even more luminous than either of these two classes, although this is not so easy a task as for the Cepheids for example whose light variations are so characteristic that they are readily recognisable.

The nearest of the large spiral nebulae is Messier 31, the great nebula in Andromeda, just visible to the naked eye on a clear, moonless night as a hazy patch of light very close to the star ν Andromedae. Nearer galaxies do exist however, the nearest in the northern hemisphere being a faint, recently discovered system in the constellation of Draco. The closest to our own galaxy however are the Magellanic Clouds, already mentioned, visible only from the southern hemisphere and distant by about 180,000 light years. At present there is much speculation among astronomers concerning these two star systems, some believing that they are merely satellites of our own galaxy and others that they should be classed as separate systems. One thing does appear to be certain however; these two systems are physically connected and actually rotating about each other, very similar to the component stars of an Algol binary.

Their importance in the present context is that, being the closest of all external systems, it is here that we would expect to discover semi-regular variables in order to compare such stars with those which we have already discussed. In recent years, several semi-regular variables have been found in the two Magellanic Clouds (Tables 10 and 11) although owing to their nature it has been possible to obtain mean cycles for only a comparatively small number of them. One point of interest which has arisen from this work is that the frequency distribution appears to reach a maximum at a mean cycle of about 100 days, confirming the earlier determinations made on similar stars in our own galaxy.

As we also know the distances of the Magellanic Clouds from the RR Lyrae stars they contain and since their dimensions are small compared with their distance from us, we may consider all of the stars contained in them to be at the same distance away from us and use this fact to obtain the absolute magnitudes of the semi-regular variables. As we can see from Tables 10 and 11, these lie between −2.0 and −5.0 indicating that in this respect there is little difference between the semi-regulars in these stars systems and those in our own.

So far then there would seem to be no great difference among the semi-regular stars of the Magellanic Clouds and those of the galaxy. When we extend our observations further however we do find one curious difference. The brightest of these stars so far discovered, actually in Messier 31 in Andromeda, has an absolute photographic magnitude of −7.5 and many others are known exceeding −6.0. It would thus appear that the stars in those extragalactic nebulae where we are able to distinguish them have a far higher luminosity than those in our own system. Quite naturally, of course, we would anticipate that in these very distant stellar systems, the brighter stars of this class will be the first to be discovered but the fact still remains that such abnormally bright semi-regular variables have never been observed in our own galaxy and it appears virtually certain that they do not exist. At present it is a puzzling problem as to why there should be this wide difference in luminosity.

Table 10
SEMI-REGULAR VARIABLES IN THE SMALL MAGELLANIC CLOUD

Star	Maximum	Minimum	Amplitude	Period (days)	M*max	M*min
HV						
829	12.7	13.5	0.8	88.5	−4.4	−3.6
834	12.3	13.4	1.1	73.5	−4.8	−3.7
1455	14.3	15.1	0.8	—	−2.8	−2.0
1877	13.8	14.4	0.6	49.6	−3.3	−2.7
1880	14.5	15.2	0.7	—	−2.6	−1.9
2084	13.6	15.0	1.4	—	−3.5	−2.1
2232	14.2	15.2	1.0	—	−2.9	−1.9

* Absolute photographic magnitude

Table 11
SEMI-REGULAR VARIABLES IN THE LARGE MAGELLANIC CLOUD

Star	Maximum	Minimum	Amplitude	Period (days)	M*max	M*min
HV						
888	13.2	14.8	1.6	—	−3.9	−2.3
916	14.0	16.0	2.0	—	−3.1	−1.1
2602	14.6	15.6	1.0	—	−2.5	−1.5
2677	15.2	16.8	1.6	—	−1.9	−0.3
2700	14.0	16.0	2.0	—	−3.1	−1.1
2740	15.0	16.8	1.8	—	−2.1	−0.3
2827	12.5	14.3	1.8	—	−4.6	−2.8
5891	13.7	14.6	0.9	—	−3.4	−2.5
5961	13.8	14.6	0.8	—	−3.3	−2.5
5964	13.5	14.3	0.8	—	−3.6	−2.8
2447	12.5	14.0	1.5	118.5	−4.6	−3.1
900	13.0	14.5	1.5	47.5	−4.1	−2.6
953	12.3	13.7	1.4	47.7	−4.8	−3.4

* Absolute photographic magnitude

In this connection, however, we must make mention of one particular variable, namely S Doradus in the Large Magellanic Cloud, an eclipsing binary of the Algol type with a period of 14,670 days. The interesting feature about this star is not its Algol-type variation, but the fact that one component turns out to be the most luminous star known (apart from novae and supernovae at maximum), having an absolute magnitude of −8.9, or about one million times the luminosity of the sun. Strictly speaking, this component is not a true semi-regular variable but belongs to a small group of variables known as the P Cygni stars in which there appears to be a continuous ejection of matter from the star in the form of an expanding shell. As might be expected, the spectrum of this star is decidedly peculiar and it will be discussed in more detail when we come to consider the eruptive variables in a later chapter.

Statistical Investigations of Semi-Regular Variables

Perhaps more than any other type of variable star, the semi-regulars have, by their very nature, provided the astronomers with the challenge of attempting to prove the existence of mathematical relationships between the observed characteristics of the light curves, spectra and galactic distribution. It is well known that a definite correlation exists between the amplitude and the logarithm of the period for the long period variables and Cepheids and for the latter, at least, the period-luminosity law is firmly established. The problem is somewhat more difficult in the case of the semi-regular stars since here the characteristics with which we are concerned are far less definitely definable. Yet in spite of this, many attempts have been made to find some correlation between the periods, amplitudes, apparent magnitude at maximum and minimum, spectral indices and galactic distribution.

Palmer has shown that there is a very small increase in the amplitude with an increase in the logarithm of the period and that statistically the amplitudes may be regarded as constant. This is a feature shown by no other class of variable star.

Turning to a possible relationship between the spectrum and the logarithm of the period, a general relation was first established by Shapley for all classes of variable as long ago as

1928, namely that in general the longer the period, the later is the spectral class. The semi-regular and irregular variables however did not fit this relationship since these stars cover the whole range of spectral types and the whole series of periods. This being so, it is possible that an independent spectrum-period relationship may hold for these variables. From a study of over one hundred stars, Palmer has concluded that there is a definite correlation between these two parameters, the longer the period, the redder is the spectrum.

When we come to consider the behaviour of these variables at maximum and minimum apparent brightness we find that the dispersion at these two extreme phases is virtually the same. Now this is quite a significant difference between the semi-regulars and the long period variables for it has been shown for the latter that the individual maxima for a particular star vary appreciably whereas the minima show very little variation.

Finally, the relationship of galactic distribution to spectrum remains to be discussed. Here some very interesting results have emerged. As we have seen, the semi-regular variables are found among almost every spectral class although with a marked preponderance towards types M and N. The variables of spectral classes B and A lie very near to the plane of the galaxy. The M stars on the other hand are found in all galactic latitudes while the N variables occupy a somewhat intermediate position, being in general nearer to the galactic plane, a situation which indicates that of the two types, the N stars have the larger masses.

12. RV Tauri Variables

The comparative complexity of the semi-regular variables is due to the fact that as a class it contains such dissimilar stars that at sometime in the future, further assessment of the position is inevitable. In the ten years between the first and second editions of the General Catalogue of Variable Stars, many of the original classifications were amended and where recent research indicated it was desirable, several stars were transferred from one class to another.

One group of variables where some confusion did, and still does, exist, are the RV Tauri stars. In particular, as we have already seen, there are several points of similarity between them and the SRd variables and as more information comes to light there will undoubtedly be further adjustments among these stars.

The RV Tauri stars are a comparatively rare class of variables, very luminous, and characterised by light curves which show alternate deep and shallow minima although interchange of minima is known resulting in two shallow or two deep minima occurring in succession. Such interchanges are observed quite frequently for some of these stars while for others, e.g. R Scuti, one of the brightest members of the class, there are sometimes prolonged periods of highly irregular activity making it quite difficult to define the alternation of minima.

The periods of these variables, which are generally defined as the interval between successive deep minima, lie between 40 and 220 days and all are of spectral classes F, G or K but with certain characteristics approaching type M3 around the time of a deep minimum. At such periods, the broad absorption bands of titanium oxide make their appearance in the spectrum.

Originally, these stars were divided into two broad groups based upon the various features of their light curves, in particular upon their mean brightness.

RVa: A group of variables, typified by R Scuti, in which the mean magnitude remains essentially constant from one cycle to the next and

Table 12

RV TAURI VARIABLES

Star	Maximum	Minimum	Period (days)	Spectrum
SZ Aqr	10.1	12.3	108.5	K0 – M3e
AD Aql	10.9	12.1	65.4	Fp
TW Cam	10.4	11.5	85.6	G0
UY CMa	10.8	12.8	113.9	Fp
RX Cap	11.6	13.7	67.9	F8 – G0
EQ Cas	11.9	13.6	58.4	Fp
V360 Cyg	10.8	13.3	70.5	Fp – G0
V399 Cyg	12.9	14.3	120.0	G2 – G8
CU Del	13.2	14.9	127.0	Fp?
SS Gem	9.2	10.7	89.3	F8 – G0
AC Her	7.1	9.4	75.5	Fp
EG Lyr	11.1	13.1	136.7	gM5
EP Lyr	10.2	11.6	83.4	Fp
U Mon	6.8	8.5	92.3	G2 – G8
TT Oph	9.9	11.7	61.1	G5 – K0
TX Oph	9.8	12.1	135.0	G0 – G8
V564 Oph	10.6	11.5	70.6	G8 – K2
EI Peg	10.0	11.5	61.2	gM5
R Sge	8.9	11.5	70.6	G0 – K0
AR Sgr	9.6	11.5	87.9	F9
AZ Sgr	11.2	12.8	114.0	F2
V760 Sgr	12.0	13.6	45.3	Fp
R Sct	5.0	8.4	144.0	G8 – M3
RV Tau	9.8	13.3	78.7	G2 – M2
V Vul	9.0	11.1	75.7	G2 – K0

All of the above magnitudes are photographic

RVb: Stars like RV Tauri and SX Centauri in which the mean brightness is found to vary over a long, secondary period.

We shall have more to say about these two classes later in the chapter. In the meantime, we begin with a more recent classification of the RV Tauri stars based upon both spectroscopic and photoelectric studies which have been carried out during the past ten years, chiefly by Preston and his colleagues. From their results they have subdivided these stars into three spectroscopic groups which may be regarded as the first real step towards a better understanding of these very interesting variables. Certainly there still remains a pronounced lack of homogeneity among these stars and many of them have been erroneously placed among the semi-regular variables at times before detailed light curves and spectra became available.

Gerasimovic was one of the first to treat the RV Tauri variables separately from the semi-regulars, showing that unlike the latter, the RV Tauri stars do satisfy the relationship between period and spectrum first enunciated by Shapley in 1928, thereby showing that in this respect they are more like the Cepheids and long period variables than the semi-regulars.

Table 12 summarises the available data on some of the RV Tauri stars which have been definitely assigned to this class. With the exception of RV Tauri itself, these are stars in which the mean brightness remains essentially constant, i.e. the RVa type of variable.

Spectral Classification of the RV Tauri Stars

Before going on to discuss the light curves of these variables, we shall consider the general spectral characteristics of the three groups into which Preston and others have subdivided them since these have a direct bearing upon what comes later in the chapter. These classes are as follows:

CLASS A: Stars of the later spectral classes G or K although certain peculiarities are evident particularly in the intensity of the CN band and as with many of these variables the broad absorption bands which often appear around deep minimum are those due to titanium oxide. So far, no bands due to zirconium oxide have been noted in the spectra of RV Tauri stars. It must also be remarked that although the absorption bands appear around the time of a deep minimum they have also been found during the preceding shallow minimum. R Scuti is typical of this class. At a deep minimum when the star fades below eighth magnitude, the titanium oxide absorption is at least as intense as that of an M3 supergiant star.

CLASS B: These are the stars which have a spectrum at maximum of type Fp but around minimum the absorption due to carbon compounds appears producing features which are very reminiscent of the rare R type spectra. We therefore have here a situation similar to the long period variables where the class may be broadly divided into the titanium and zirconium oxide stars or the carbon stars.

The attendant changes in the intensity of the absorption bands due to carbon compounds, in particular CH and CN, with light phase are very conspicuous in this class. It must be pointed out however that the spectra are highly peculiar at all phases of the light variation. For example, around maximum, the spectral class as estimated from the absorption lines of hydrogen corresponds to a range from F5 to GO sometimes even later, whereas that found from the lines of singly ionised calcium is invariably earlier. We have come across this anomaly before in the case of the RR Lyrae variables and it is quite possible that the same explanation applies to this class of RV Tauri stars. Provided the spectrograms have been sufficiently well exposed, stars of Class B show totally different characteristics at the violet end of the spectrum (where as we will recall the carbon absorption bands occur) to those of the preceding class.

CLASS C: These are variables whose spectra are very like those of the previous class, differing in the absence or the general weakness of the carbon absorption. Because of this likeness, it is sometimes quite difficult to differentiate between these two classes and a detailed examination of the violet portion of the spectrum, under sufficiently high dispersion, is necessary for accurate classification.

One must not imagine however that every RV Tauri variable which has been examined spectroscopically can be placed unambiguously into one or other of the above classes. Many have qualities intermediate between classes B and C. Then, too, there are several peculiar stars which show marked similarities to the red Me stars and in spite of the fact that their light curves would define them as RV Tauri variables, it is quite possible that as our knowledge of these stars increases it will be found that they are not really members of this class.

Light Variations of RV Tauri Stars

These stars form such a heterogeneous group that it is hardly surprising we find a wide variety of light curves and for a full

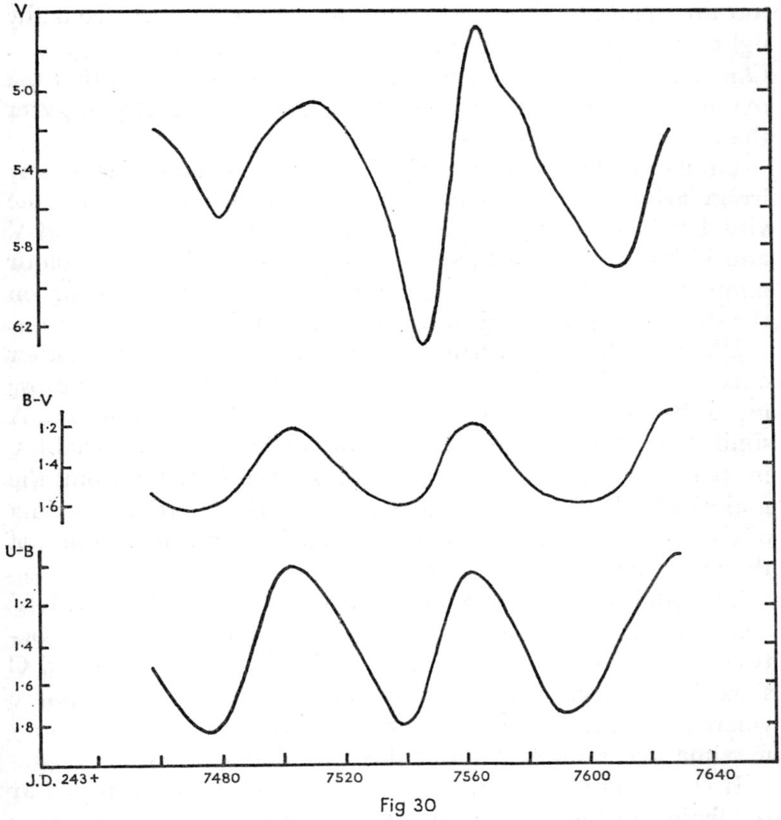

Fig. 30 – Three colour photometry of R Scuti (*Astrophysical Journal by kind permission of the University of Chicago Press*).

discussion of these the reader may be referred to the excellent and comprehensive study of these stars made by Gaposchkin, Brenton and Gaposchkin given in Harvard Annals, Volume 113, No. 1. Most of these stars are supergiants and all are highly luminous, several being visually bright objects which may be followed throughout the whole of their light cycles by means of very moderate instruments; indeed, AC Herculis, U Monocerotis and R Scuti may be covered with binoculars except during a deep minimum of the first.

Not only do the minima vary greatly in depth, the maxima too are generally far from constant as may be seen from the light curves of R Scuti (Fig. 30) and AC Herculis (Fig. 31). One feature which is common to many of these stars is that the maximum which follows a deep minimum is generally brighter than that which precedes it.

Unlike the long period variables there is seldom, if ever, any irregularity on either the ascending or descending branch of the visual light curve although such humps are found in the B-V and U-B curves when these stars are examined by three-colour photometry. Usually, though not invariably, these occur on the descending branch just before a deep minimum.

We may also draw certain conclusions from the three-colour curves; thus the minima in the U-B curves occur before those in the B-V curves which in turn precede the visual minima. A similar effect is found for the maxima. A few of these variables, in particular U Monocerotis and R Scuti differ from the majority in that these deviations are so marked that the maxima of the U-B and B-V curves almost coincide with the minima of the visual light curve.

This shift in phase, especially between the visual and the B-V colour index curves, appears to indicate a real difference between the Class A stars and those of Preston's Classes B and C since in the former the shift is almost independent of phase whereas in the latter it is appreciably smaller at primary maxima and minima than at the secondary phases.

It is very important also to apply corrections for interstellar reddening to the colour indices of these variables since by far the greater number of them lie more than 3,000 light years distant from us.

Spectral Variations of R Scuti

Much of our present knowledge concerning the physical characteristics of the RV Tauri stars has come from the various spectroscopic studies which have been carried out on R Scuti,

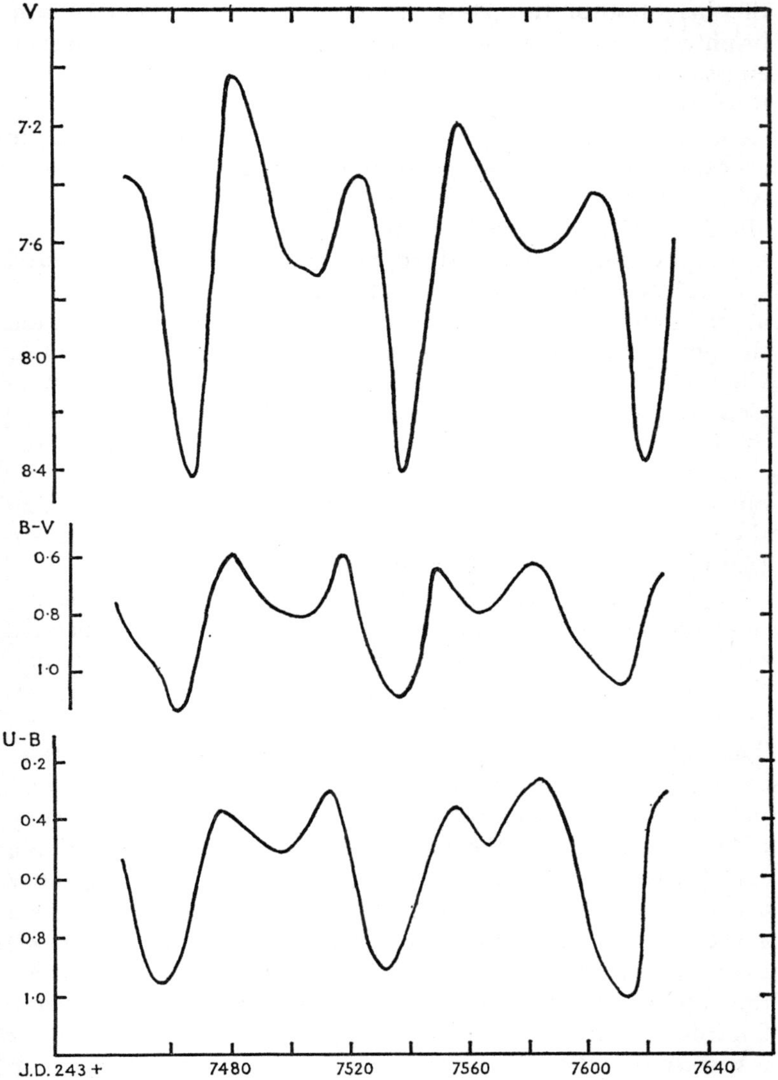

Fig. 31 – Three colour photometry of AC Herculis (*Astrophysical Journal by kind permission of the University of Chicago Press*).

the brightest of this class of variable and incidentally the brightest of the intrinsically variable stars in which hydrogen emission and line doubling occur around the phase of maximum brightness. One of the many peculiarities of a great number of these stars is the appearance of the well known titanium oxide absorption bands in the spectrum during the deep minima. It is very tempting indeed to presume that such stars are closely allied to the long period variables or the semi-regulars which have previously been described, assuming too that the observed spectral differences are due to the higher surface temperature at maximum with a consequent wider temperature variation throughout the light cycle. It has already been pointed out in Chapter 10 that a few of the long period variables, e.g. R Centauri, have light curves very like the RV Tauri stars owing to the pronounced hump on the descending branch of the curve which gives the appearance of a secondary minimum. Such a relationship was indeed considered quite possible during the early investigations of these stars but later work has shown that the true picture is of a far more complex nature.

High dispersion spectrograms taken with large instruments have shown that all of the strong metallic absorption lines are double at maximum and furthermore, between the two absorption components there is a weak, but clearly identifiable, emission line which rises above the background continuum. This phenomenon of line doubling accompanied by the appearance of the bright hydrogen emission lines so complicates the spectrum of R Scuti at maximum light that it is extremely difficult to compare the variable with normal, non-variable stars having similar spectra. As we have seen it is now generally accepted that these stars are nearly all supergiants and from a comparison of the relative abundance of neutral and ionised metals present it would appear that they most closely resemble the supergiant Ko stars at maximum. We must not however fall into the error of taking this resemblance too far; at most it is merely superficial since there are many discrepancies in the spectra of the RV Tauri variables, principally in the weakness of the emission characteristics and also the general weakness of the absorption lines, especially those of the rare earth metals.

During the rise to maximum, the hydrogen lines begin to

appear in emission in the form of bright borders on either side of the absorption lines, the emission border on the violet side being the stronger of the two, their intensities not being constant but varying from one maximum to the next. The origin of the emission lines within the star is not known with certainty but from the results which have so far been obtained it is thought to be within the stellar atmosphere.

We now come to a consideration of the broad absorption bands of titanium oxide which are found in the spectra of many of these stars when they pass through a deep minimum. From our discussion of the long period and semi-regular variables, these are characteristic of the low temperature at minimum, of the order of 1,800°K which allows the formation of these relatively stable molecules within the outermost layers of the star.

Are the temperatures prevailing in the atmospheres of the supergiants of the RV Tauri class sufficiently low to provide similar physical conditions? Clearly this is an important point if we are to devise a satisfactory model for these stars. Even the early low-dispersion spectrograms taken at minimum light were sufficient to show that they are not. In view too of the fact that the subordinate line intensities are not reconcilable with those found in typical M type variables, we must look for another explanation.

Here it will be worthwhile to consider briefly the factors which influence the dissociation of such molecules into their component atoms. The heat of dissociation is one of the most important factors but since this is constant for any given molecule, such as titanium oxide, the two major parameters are the temperature and the pressure; the higher the temperature and the lower the pressure, the greater the percentage of the molecules which become dissociated. In this connection, it is important to note that titanium oxide is much more sensitive to pressure changes than zirconium oxide and at a pressure of 10^{-6} atmosphere is completely dissociated at 3,000°K whereas under similar conditions zirconium oxide is only about sixty per cent dissociated. Since, in the atmosphere of any variable star, the temperature and pressure almost certainly vary simultaneously, the effects produced in the band spectrum are necessarily very complex.

Now it seems highly unlikely that the surface temperatures of these stars at minimum is sufficiently low to allow a fairly high proportion of the titanium oxide to remain undissociated and at present there is no conclusive evidence for the suggestion that the titanium oxide bands originate in a gaseous envelope surrounding the star although this possibility is now being seriously investigated by many astrophysicists. Such a circumstellar envelope is not unfeasible. A similar idea has been

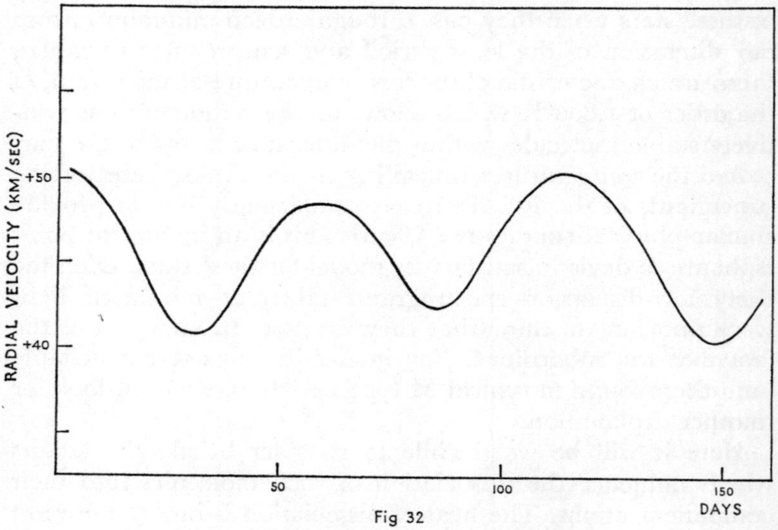

Fig. 32 – Radial velocity curve of R Scuti.

put forward to explain the unpredictable deep minima of the R Coronae Borealis variables where it appears that the condensation of carbon at some distance from the star may be responsible. Certainly, if this is so, we may postulate a reasonably low temperature at some comparatively great distance from the surface of an RV Tauri star.

Radial Velocities

The radial velocity curves for R Scuti have been obtained by several astronomers and as they provide us with important, though not absolute, evidence for the existence of a large circumstellar envelope around R Scuti, it is appropriate at this point to consider them in some detail (Fig. 32).

The surprising result which has emerged from these spectroscopic investigations is that two different radial velocity curves may be drawn from the lines of neutral iron in the spectrum and to understand the reason for this we must first discuss the term *excitation potential.* We have already learned that atoms may be ionised, that is one or more of their outer electrons may be removed so that the atom is no longer electrically neutral but becomes positively charged. This may be brought about either by a sufficiently violent collision with another particle or by the absorption of a quantum of very short wavelength having the requisite energy to overcome the attraction between the electrons concerned and the atomic nucleus. Under terrestrial conditions such ionised atoms have extremely short lifetimes since there is always a plentiful supply of electrons available to recombine with these ions, but in the highly rarified atmospheres of these supergiant stars, they are capable of existence for comparatively long periods.

Suppose however that the atom absorbs a quantum of longer wavelength in which the energy is not quite enough to remove an electron altogether from one of the outer orbits. What happens now is that an electron jumps into a larger, outer orbit around the nucleus and the atom is said to be excited. It will then contain a greater and greater amount of energy as the excitation increases but will, unlike an ion, remain electrically neutral. The amount of energy contained within such an atom is known as the excitation potential and we conveniently measure this in electron volts. Where the excitation potential is very low, less than about 0.2 electron volts, we say that the atom is in its ground state.

In the case of R Scuti, and several other RV Tauri variables, two sets of neutral iron lines are present in the spectrum, the subordinate lines with an excitation potential greater than 0.9 electron volts and the ground state lines in which it is less than 0.2 electron volts.

From Fig. 32 it will readily be seen that the two radial velocity curves are radically different; not only is the amplitude smaller in the latter case, but the curve obtained from the ground state lines does not follow that of the subordinate lines. From this and the fact that the ground state lines are found to be abnormally strong compared with the remaining absorption lines of neutral iron, it appears highly probable that they originate in a circumstellar envelope where we would expect the particles to have a lower energy than those within the stellar atmosphere.

Origin of the Light Variations in RV Tauri Variables

The question why these stars should exhibit such curious light variations which are of a far more periodic and regular nature than those found in the semi-regular variables, being more like the long period variables in this respect, is not an easy one to answer and as yet no fully satisfactory explanation has been forthcoming. Many attempts have been made to explain the semi-periodic light changes on the supposition that these stars are pulsating variables which are undergoing an overtone pulsation in their atmospheres. Certainly from a consideration of the relationship between the light and radial velocity curves, all of the evidence points to some form of pulsatory mechanism. As with the semi-regular variables any detailed analysis of the light curve in order to derive the existence of two or more periods is a long and laborious task, requiring a continuous light curve covering many years and here we come up against one of the major difficulties. Such light curves without any gaps are available for only one or two stars; only R Scuti and AC Herculis fulfil this condition to any degree.

For these two variables, theoretical and observational calculations suggest that there are two basic periods – the 'classical' period which is that normally obtained from the light curve being the interval from one deep minimum to the next and an 'essential' period which is half of the 'classical' period and represents the first overtone of a hypothetical Cepheid-like fundamental. On this basis, whether the periods of these stars, as usually defined, are true periods in the strict sense of the word is debatable.

Obviously, whatever the basic cause of the RV Tauri-like light variations may be, it would seem to be an unstable one. This is well borne out by the rarity of these stars when compared with the large numbers of long period variables and Cepheids which are known. It may, of course, be argued that this rarity is illusory and that many more still remain to be discovered in the galaxy but this is unlikely for two reasons. Firstly, as we have seen, they are all very luminous stars visible for great distances and secondly their light curves are quite distinctive with large amplitudes. A further point in favour of the RV Tauri phenomenon being due to stars at a critical stage of their evolution was pointed out by Gerasimovic almost forty years ago when he noted that the periods of these variables fall into one of the minima of the period-frequency curve for those stars which are intrinsically variable. At the present state of our knowledge therefore it appears they are stars of high mass which have evolved far to the right of the main sequence and in so doing, cross a very narrow point of instability in which some form of pulsation occurs. The extreme rarity of these variables can then be ascribed to the basic rarity of such massive stars and to the short duration of this unstable state compared with the lifetime of such a star.

Masses of RV Tauri Variables

We cannot measure the masses of these variables directly, of course, for the simple reason that they are single stars and we are therefore forced to rely on more indirect methods. Being supergiants these stars have clearly evolved far to the right of the main sequence becoming larger and brighter in the process until they are now among the most luminous stars known having absolute visual magnitudes of between −4·5 and −5·0.

We may therefore tentatively infer their masses from their position on the Russell-Hertzsprung diagram. These turn out to lie between 20 and 25 times that of the sun, truly very massive bodies.

It is possible also to obtain a similar value from the mass-luminosity relationship although it must be borne in mind here that, strictly speaking, this law holds only for stars on or very close to the main sequence. With this qualification in mind, we see that the RV Tauri stars are among the most massive variables known.

SX Centauri Variables

These are the RVb variables mentioned at the beginning of this chapter with RV Tauri itself a representative of the group. The distinction between these and the RVa stars is extremely nebulous and in a great many cases the only difference lies in the existence of a well-developed secondary period which is much more pronounced than any so far found for stars of the RVa class.

The light curves of the SX Centauri variables are very

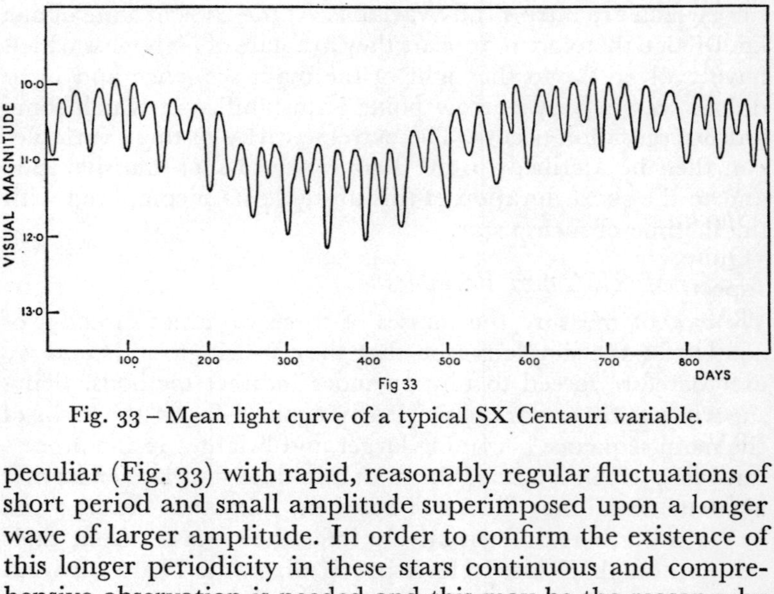

Fig. 33 – Mean light curve of a typical SX Centauri variable.

peculiar (Fig. 33) with rapid, reasonably regular fluctuations of short period and small amplitude superimposed upon a longer wave of larger amplitude. In order to confirm the existence of this longer periodicity in these stars continuous and comprehensive observation is needed and this may be the reason why so few can be placed in this particular class at the present time.

The periods of the more rapid fluctuations range from 30 to 70 days while the secondary periods are of the order of 700 days with the exception of IW Carinae which is given as 1,500 days although there are several peculiarities associated with this particular variable, especially in its spectral characteristics which indicate that it may be a binary system. In Table 13 are given details of this small group of stars but here again it must be emphasised that their distinction from the RVa variables is very vague. RV Tauri which has been included in Table 12 has been shown to possess a fairly well defined secondary period of approximately 1,230 days.

Table 13
SX CENTAURI VARIABLES

Star	Maximum	Minimum	Primary Period (days)	Secondary Period (days)	Spectrum
IW Car	7.90	9.62	67.5	1500	G0+A0
SX Cen	9.92	12.4	32.8	620	F5
DF Cyg	10.80	15.2	49.8	728	cG5+cK4
SU Gem	9.90	12.2	50.1	690	F5–G6
AI Sco	9.40	12.6	71.0	960	G0–K2

All of the above magnitudes are photographic

Spectra of SX Centauri Variables

Quite clearly, with the single exception of IW Carinae, the spectra of this small class of stars are quite similar to those of the RVa group of RV Tauri variables.

The spectrum of IW Carinae at certain phases of the light curve exhibits some of the features of an early A type supergiant (such features have also been reported in the spectrum of the RVa type variable EP Lyrae) although there is no firm evidence at all to lead us to suppose that IW Carinae is an eclipsing system.

13. Irregular Variables

Under this heading are gathered for convenience a large number of stars which have, unfortunately, been only poorly observed in the past compared with most other kinds of variable stars. In the foregoing chapters we have seen how, even with the semi-regular variables, some degree of regularity still remains in the light variations which enables us to find a mean cycle for the star although in most cases this may be submerged beneath a complex of several periods which can be separated with more or less difficulty only from a long and detailed investigation of the light curve. The general behaviour of the long period variables for example is far less regular than the high precision of the Cepheids or the eclipsing binaries but even here the form of the light curve is sufficient to define a mean period. The semi-regulars and others such as the RV Tauri stars are characterised by a certain periodicity on which may be superimposed various irregularities and finally the SX Centauri variables, formerly supposed to be completely irregular, have shown two periods to be present in their light variations.

We must now take a look at quite a large group of stars in which all trace of periodicity has vanished. These are the true irregular variables and we shall see that most, if not all, of them are cool, red giants with late type spectra showing very little evidence of emission lines. This latter feature must be treated with caution however since these stars, as a class, have been far less assiduously observed than the majority of other similar variables and it is perhaps presumptuous to be dogmatic about them.

The amplitudes of these stars are all small, seldom exceeding two magnitudes and although several hundred are known, the light curves for only a few of them have been determined for any length of time. With increasing observational material it may well be that some of them will be found to belong to the semi-regular variables since, as has been shown earlier, it is very difficult to discover the underlying periods of most of these stars unless they are comprehensively observed over a long number of years. The light curve of V Aquilae, one of the best observed of this class, is shown in Fig. 34.

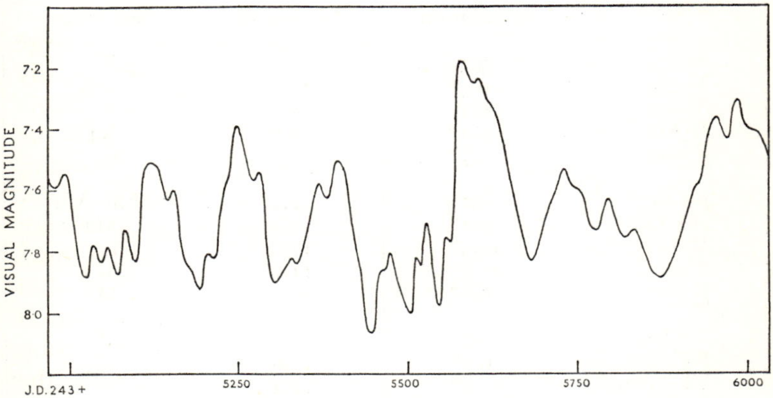

Fig. 34 – Light curve of V Aquilae.

One of the brightest of this class of variable star is the fourth magnitude ρ Cassiopeiae which was discovered to be variable by Miss Wells in 1901 and was classified as irregular in 1907 by Pickering and Cannon on the basis of their observations during the previous four years. The light curve of this particular variable is peculiar in the extreme consisting of slow waves of varying amplitude and characterised at long, irregular intervals by fairly deep minima. The presence of these deep minima in the light curve has led to it being sometimes classed as an R Coronae Borealis variable of small amplitude but this is no longer thought to be the case particularly as the spectrum does not resemble that of these stars.

One marked pecularity of the light curve of ρ Cassiopeiae is the apparent existence of three fairly distinct modes with quite abrupt changes occurring from one to the next. Such changes have been observed in 1911, 1922 and again as recently as 1946 when a very deep minimum occurred which was well followed both visually and photoelectrically. Within each of these separate modes there does seem to be a certain kind of semi-regularity, what we may term a mean cycle, this amounting to

Table 14
IRREGULAR VARIABLES

Star	Maximum	Minimum	Spectrum
U Ant	5.7	6.8	N
V Aps	10.5*	11.5	Mb
V Aql	6.7	8.2	N
V Ari	7.7*	8.3	R4
S Cae	9.7*	10.5	K
ρ Cas	4.1	6.2	cF8 – M2
XY Cen	9.4*	10.4	M2
W CrA	11.3*	11.8	R3
SS CrB	10.0*	11.2	M5
R Crt	9.8*	10.4	M4
TZ Cyg	9.6	11.7	M7
RR Eri	9.0*	10.0	M5
TU Gem	11.2*	12.4	N
π^1 Gru	5.8	6.4	S
SX Lac	9.6*	10.4	K2
Z Lup	10.5*	11.6	N
X Lyr	8.8	9.5	M4
S Men	10.2*	11.5	Mb
Z Mon	9.0*	10.1	K5
X Nor	12.5*	13.5	N
RT Ori	8.1	8.9	N
T Per	8.3	9.3	cM1
W Pic	11.8*	13.2	N
S Sct	7.0	8.0	N
Y UMa	9.5*	11.2	M7

* Photographic magnitude

about 450 days for the first mode, 200 days for the second and approximately 300 days for the third.

The spectral variations of this supergiant star have been both wide and varied since its light variability was discovered at the beginning of the century. The spectrum is normally of type F8 indicating a surface temperature of about 6,500°K. Usually, the intense lines of ionised calcium are visible in the violet and ultra-violet together with numerous absorption lines of the metals. The only peculiarity which has been recorded for the F type spectrum of this star is the marked weakness of the Balmer lines due to hydrogen during the latter half of the 1950s. That this is due to a filling in of these lines by emission lines has been shown by Sargent, proving that there is not a deficiency of hydrogen in this particular variable.

Prior to 1930, the spectrum at times showed the characteristics of a K star with metallic lines of iron being quite prominent and during the deep minimum of 1946 the absorption bands of titanium oxide appeared giving a spectrum resembling that of an M type star. As the brightness increased again following the minimum, the spectrum of ρ Cassiopeiae gradually began to revert to the earlier type F but between 1952 and 1957 further changes became evident. On several high-dispersion spectrograms many of the stronger lines appeared double due to the presence of another line towards the violet of the normal one. All of this points to the presence of an expanding shell of gas around the star and it seems highly probable that the titanium oxide absorption bands observed during the minimum originated within this shell of cooler gases. Measurements show that the shell is expanding outward from the star at about 40 kilometres per second and although this is less than the escape velocity for a supergiant star such as ρ Cassiopeiae, there is no evidence in the spectrum to show that any of the ejected material is actually falling back towards the star's surface. Evidently, in spite of this low velocity it is escaping outwards into space.

Although there is evidence that most of the M type supergiant stars possess similar circumstellar envelopes, it is most unusual for these to occur with supergiants of early spectral types, e.g. F8, and only a few such stars are known apart from ρ Cassiopeiae. The star 89 Herculis is one such star and possibly also ρ Pegasi which is also variable.

The distance of ρ Cassiopeiae has been found to be about 6,500 light years. Its diameter is believed to be 700 times that of the sun and its mass approximately twenty-five times greater than the sun. Such a supergiant star must only recently have begun to evolve to the right of the main sequence and although the loss of mass by ejection into the circumstellar envelope is quite clearly of a sporadic nature, it is quite conceivable that the mean mass loss is comparable with that which we would expect due to its evolution.

A further small group of irregular variables which we must consider are the Be stars of which the brightest example is γ Cassiopeiae, normally a star of magnitude 2.25 being readily visible to the naked eye. The members of this group do not all show the same characteristics but the basic phenomena are very similar and γ Cassiopeiae may be taken as representative of this group of variables.

Since several of these stars are bright objects they were among the first to be observed spectroscopically and consequently much is now known of their light changes and the variations which take place in their spectra. As to a detailed explanation of the basic causes of these variations, the position is still far from satisfactory. It does appear that the bright emission lines in the spectra of these B type stars originate either in a shell or in a vastly extended atmosphere around the star, or in some cases there may even be a rotating disc formed by the ejection of mass from the star by virtue of its extremely rapid axial rotation. The absorption lines in the spectra of these variables are generally broad and diffuse, indicative of such high rotational velocities.

From the light curve of γ Cassiopeiae (Fig. 35), the irregular nature of its variation is readily seen.

Much of our present knowledge of these stars has undoubtedly been derived from an examination of the spectral rather than the light changes and it is worthwhile here to consider these in detail with particular reference to γ Cassiopeiae, not only because this is the brightest and best observed of these variables, but also because this star has shown all of the spectral changes associated with variables of this class together with others which are possibly unique.

The earliest spectroscopic observers noticed small changes in the intensity of the emission lines which at times have been regarded as spurious but which are now thought to have been real. In general however the star appears to have been fairly quiescent until about 1929 with the brightness varying by 0.2 magnitude at the most. During 1932, however, the violet component of the double emission lines began to increase rapidly in intensity as also did the Balmer emission lines, several of the higher members becoming visible for the first time. By the following year, some of the neutral helium lines were also showing in emission. Late in 1935 there were further developments, in particular the line of neutral helium appeared

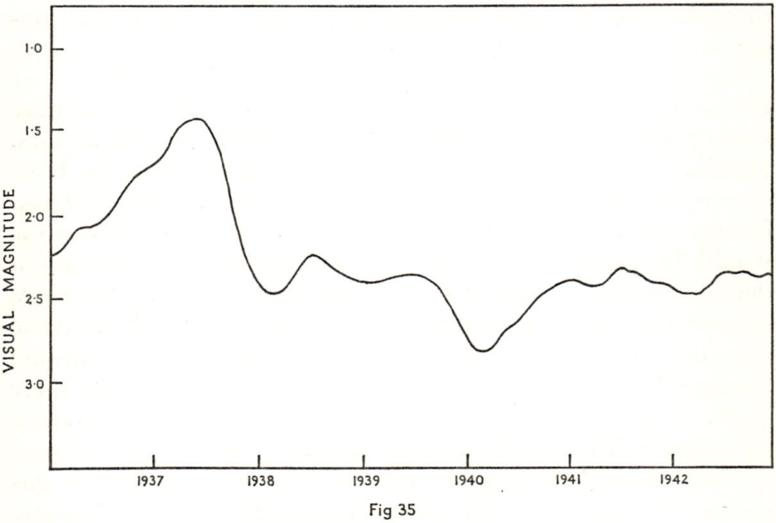

Fig. 35 – Light curve of γ Cassiopeiae.

strongly in absorption as a shell developed, reaching its maximum intensity early in 1936.

The following year the star reached its maximum visual brightness while at the same time the spectrum was decidedly peculiar, exhibiting very strong emission lines of hydrogen, helium and ionised iron and fainter emission of several metals. Following these striking variations, the star faded somewhat, reviving in brightness during 1938 and then declining once more. Further shells have developed since then, each accompanied by changes in the spectrum which have, at times, been extremely rapid.

Several attempts have been made to correlate the changes in the spectrum with the light variations but although they do appear to be related during certain phases, this relationship is not maintained over the whole of the period since 1929. One interesting feature of the spectrum during the major outburst of 1937 is that the emission lines were all very sharp, showing that the circumstellar envelope giving rise to them had a fairly slow period of rotation. If we accept this as a true interpretation of the facts then the shell must have been at some considerable distance from the photosphere of the star owing to the conservation of angular momentum.

As the shell dissipated, the star became appreciably bluer than before, the colour index indicating a temperature of more than 30,000°K which is higher than normal for type B stars. Detailed measurements of the underlying continuum have shown however that this effect was brought about, not by an increase in the intensity of the blue end of the spectrum (as would be caused by a true rise in temperature) but by a decrease in the strength of the red continuum. We can therefore assume that the temperature of the photosphere did not rise after the outburst and that the change in colour index was due to some other mechanism taking place just below the outer shell. There are several difficulties in explaining the varying intensities of the double emission lines.

Undoubtedly a Doppler shift due to a rotating shell of gas contributes to these variations in intensity but additionally there must be some other mechanism occurring, possibly the existence of fairly localised, but violent, turbulences within the shell. The idea that this star might have an unseen companion which eclipsed the rotating shell has been put forward by some astronomers but there is, at present, no evidence for this from the radial velocity measurements.

Some of the other Be variables appear to be more regular than γ Cassiopeiae in their spectral variations. Many attempts to find such regularities in this particular star have been made but with only limited success. Doubtless further work on these stars will provide more information about these peculiar changes, information which could bring to light vital clues concerning the early evolution and constitution of these supergiant stars.

μ *Cephei*

Since this star was first recognised a variable, in 1848, a vast number of observations, mostly visual, has accumulated and there seems no doubt whatever that its light variations are truly irregular. Hassenstein concluded from a study of the observations up to 1938 that the primary variations have amplitudes which lie between 0.2 and 1.4 magnitudes and also that the intervals between the maxima and minima may be satisfied by a period of 730 days although he also stressed that there was a long period of slightly more than three years when the star did not appear to vary at all in brightness. Secondary fluctuations of less than 0.2 magnitude are also present in the light curve. These usually appear at times of broad maxima and have durations of several months and for these there is not even an approximate periodicity. Coupled with the above period, Hassenstein concluded that there may be two more, one of about 730 days as previously found by Tsesevitch and a much longer one of around 5,000 days. Balasoglo used the method of harmonic analysis in order to interpret the light curve, a method which has proved extremely useful in the analysis of the light curves of the semi-regular variables. Several criticisms have been levelled against this work. Since it presupposes that the harmonics will be stable, any results obtained are of doubtful value in the case of such an irregular variable and we must treat any period derived by it with considerable caution.

From the visual light curve (Fig. 36), it will be seen that the light variations of this star appear to be in the form of several fairly constant plateaux with the star brightening and fading fairly regularly from one region of almost constant brightness to another. Such a light curve is extremely difficult to interpret as a result of superimposed sine waves, nor can we explain it by postulating a series of damped oscillations produced by local disturbances on the star. Photoelectric observations by Larsson-Leander have suggested that at times there may be some slight evidence for damped oscillations of a very small amplitude, certainly not more than 0.03 magnitude but this is by no means certain.

We find that the colour of this star shows very little change with the brightness. It is a very red star of spectral class M2e, a fact which possibly drew the attention of the early investigators to an examination of its possible variability since at the beginning of the nineteenth century it was generally believed that all red stars were variable to a certain extent (a supposition which is not without some foundation). However, although the colour index alters slightly from about +2.3 to +2.5 there does not appear to be any correlation between this and the observed light changes.

Quite recently, Grigoryan has found that there are fairly

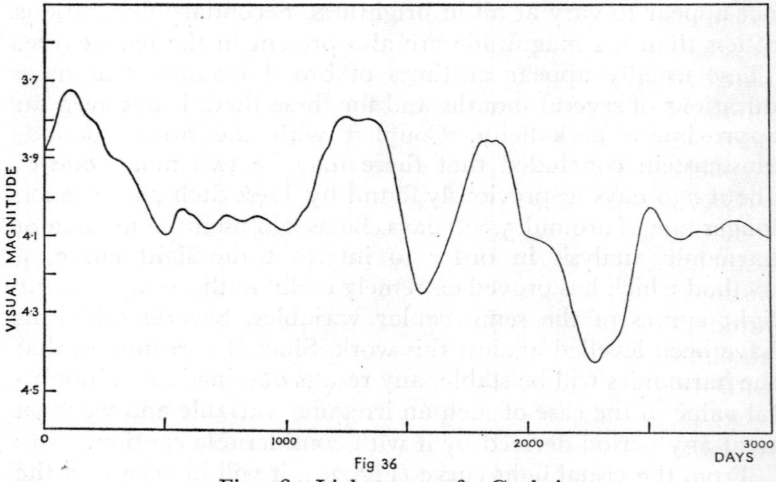

Fig. 36 – Light curve of μ Cephei.

large fluctuations in the degree of polarisation of this star but again there appears to be no correlation between this and the light variations. So far we have no explanation of the nature of these polarisation changes.

The presence of these long plateaux in the light curve when the brightness remains essentially constant suggests that there may be an optimum magnitude of about 3.99 for this star which represents a stable state of the variable. We can visualise μ Cephei as a red giant, on the surface of which there are large areas where the effective temperature differs from that over the remainder of the star. The temperature gradient is probably not steep since the magnitude variations of this star are small. Depending upon whether the effective temperature of this area is above or below the norm, the result will be a long, flat plateau which may be either brighter or fainter than this 'normal' magnitude. Kreiken has suggested a rotational period of a little over 700 days for this star although this is certainly not obvious from a study of the light curve.

Irregular Variables in Galactic Star Clouds

As we have seen in the case of the semi-regular variables, one way to obtain absolute magnitudes for these stars is to consider isolated groups in certain regions of the galaxy where it is possible, by other means, to calculate their distance and this is particularly so if there are also Cepheids within these star clouds. Indeed, the distance factor is really superfluous since once we know the absolute magnitudes of the associated Cepheids, we can obtain those of the irregular variables directly.

One of the largest of these star clouds lies within the constellation of Cygnus and contains several Cepheids together with a large number of irregular variables. From the mean distance modulus (absolute magnitude minus apparent magnitude) of the Cepheids BZ Cygni and V383 Cygni, the absolute magnitudes of many irregular variables associated with these stars has been found. These are found to vary from -3.2 to $+2.6$ and quite clearly the absolute magnitudes of the irregular variables cover a wide range. Several other similar star clouds containing both types of variable are found in Scutum, Perseus and Orion and all yield results in good agreement with those found above and by independent methods such as the parallactic motion of these stars.

Certainly the range of absolute magnitudes found for these irregular variables is large. There are, of course, several reasons for this. Because the class as a whole contains some very dissimilar stars, ranging from bright B type variables to giant and supergiant stars of late spectral types, we would anticipate a wide variation in absolute magnitudes. Of the variables considered, many are below twelfth magnitude at maximum and no satisfactory spectra are available to differentiate between these three main groups. Then too, if we consider the irregular nature of their light curves, there is the added difficulty of determining just what is their magnitude at maximum. Finally, it is possible, particularly with the fainter members, that some may actually be semi-regular variables and do not belong in this class at all. For example, the variable DP Orionis was originally classed as an irregular variable with an absolute magnitude of −4.69 which was recognised at the time as being too bright for this particular type of star. Later work, however, has shown that this star is actually an RV Tauri variable and as we have seen, these stars are among the brightest of all variables excluding the novae. Once the true nature of DP Orionis was recognised, the anomaly disappeared.

Irregular Variables in Globular Clusters

Since the distances of most of the globular clusters are quite accurately known, they would appear to be ideal regions for the determination of absolute magnitudes. Unfortunately, as with the semi-regular variables, very few irregulars have been discovered within these clusters. From those which have been found, absolute magnitudes of the order of −1.7 have been calculated, agreeing very closely with the values already determined for such stars within the galaxy.

Irregular Variables in Other Galaxies

Several irregular variables have been found in the Magellanic Clouds and the nearer extragalactic nebulae but here again it is difficult to differentiate the true irregulars from the semi-regular variables. It is almost certain that as more data is accumulated for these stars, many of those now classed as irregular will be found to have certain periodicities in their light variations which will ultimately place them in the class of semi-regular variables already discussed.

PART THREE

Eruptive Variables

14. U Geminorum Variables

One of the most remarkable and certainly most interesting classes of variable star is that typified by U Geminorum, normally an extremely faint star of about magnitude 14.4. It was first discovered by Hind as a ninth magnitude object on December 15 1855, readily distinguishable from the background stars by virtue of its pronounced bluish colour. Over the next three days it faded by half a magnitude and at the end of three weeks it had dipped below twelfth magnitude and was no longer visible in Hind's instrument. The fact that its rise had clearly been extremely rapid, coupled with its undoubted blue colour, led Hind to believe that it was a typical nova. The following March, however, it was seen again by Pogson proving beyond doubt that it was not a nova.

Hind's star is the prototype of a group of cataclysmic variables often termed the dwarf novae since they are prone to rapid, temporary outbursts during which the star brightens by as much as five magnitudes in a single night. Although the period between successive maxima is by no means constant, being in fact quite unpredictable, the duration of an average cycle may be determined for any particular U Geminorum star provided the mean of a sufficiently large number of outbursts is taken.

With several of these variables this is a relatively simple task but in the case of U Geminorum itself and also X Leonis, it is not so easy since these stars are situated close to the ecliptic and for about three months of the year, from around June to August, they are unobservable resulting in several maxima escaping observation.

More consistent results have been obtained from a study of SS Cygni, the brightest of this class, every maximum of which has been followed by the Variable Star Section of the British

Astronomical Association and the American Association of Variable Star Observers since its discovery by Wells in 1896. Averaging the interval between maxima over ten or twenty years yields a mean figure close to 50.4 days, i.e. six or seven maxima a year. For most of the time this star behaves fairly consistently but in 1960 for example there were no fewer than twelve maxima giving a mean period for that particular year of

Table 15
U GEMINORUM VARIABLES

Star	Maximum	Minimum	Period** (days)	Spectrum
VZ Aqr	11.8*	15.0	49	Pec
UU Aql	11.0*	16.8	56	Pec
SS Aur	10.5	14.5	54.1	Pec
SS Cyg	8.1	12.1	50.4	A1+dGep
EY Cyg	11.4*	15.7	40.8	Pec
U Gem	8.8	14.4	102.9	Gep?
VW Hyi	8.4*	13.4	35	Pec
X Leo	12.0	15.1	22	Pec
AY Lyr	12.6*	16.0	23	Pec
CZ Ori	11.8*	16.2	38	Pec
RU Peg	10.0	13.1	70	Pec+dG3ep
UV Per	12.3	15.2	300	Pec
UZ Ser	12.0	16.6	40	Pec
SU UMa	11.1	14.5	16	Pec
SW UMa	10.8*	16.2	1000?	Pec

* Photographic magnitude
** Mean period averaged over several cycles

only about thirty days. At other times only four or five maxima are observed in a year. It is hardly surprising therefore that such stars must be observed on every possible occasion since it is quite impossible to forecast when an outburst will occur, nor the type of maximum which will follow the rise.

At the present time only about 150 of these variables are known but their light variations are so unique that they are readily recognised. Most decidedly the visual effects produced by the abrupt increase in brilliance give the impression that they must be closely related to the true novae but the analogy must not be carried too far, nor must we attempt to draw any dogmatic conclusions simply from the light curve. The U Geminorum variables are not concentrated in the direction of the galactic centre, nor along the boundaries of the Milky Way as is the case with the novae. In addition there are no forbidden lines in their spectra as we would normally expect from expanding shells of gas thrown out by a nova-like explosion.

Although quite a large number of these variables have been discovered, very few of them are brighter than 11.0 magnitude at maximum and the large majority are fainter than 15.0 magnitude at minimum. As a result only a small proportion of the total number have been really thoroughly investigated and their mean periods determined with any degree of precision. Table 15 lists the better observed and brighter stars of this class.

Light Variations of U Geminorum Variables

Although the light curves of these variables are in many ways similar, characterised by abrupt outbursts of light and slower declines to minimum, there are certain peculiarities associated with individual stars which are worthy of mention in detail.

The type star U Geminorum, together with several other members of the class, have two distinct kinds of maxima – long and short – defined by the length of time spent at maximum brightness. In each case the rise is extremely rapid often being accomplished in a single night, the ensuing decline being much more protracted. In almost every star there is a strong tendency for the variable to attain a greater brightness at a long maximum than at a short. As a general rule, long and short maxima occur alternately but there are variations and it is not unknown for two or more of the same type to follow each other.

SS Cygni, the most completely observed of these variables, also exhibits such long and short maxima but in addition, a high proportion of 'anomalous' maxima have been observed. Such maxima are characterised by a relatively slow rise often with marked irregularities on both the ascending and descending branches of the light curve and are almost always fainter than either of the other two types. The 'anomalous' maxima also tend to be rounded at the peak rather than flat. (Fig. 37).

A great deal of work has been done on the classification of the different types of maxima for this variable, mainly by Campbell who based his findings on almost 45,000 observations of SS Cygni. Four basic types were recognised by this astronomer – A, B, C and D, the letters giving an indication of the rapidity of the rise to maximum ranging from type A, the steepest, to type D, the most protracted. To each letter is given a suffix

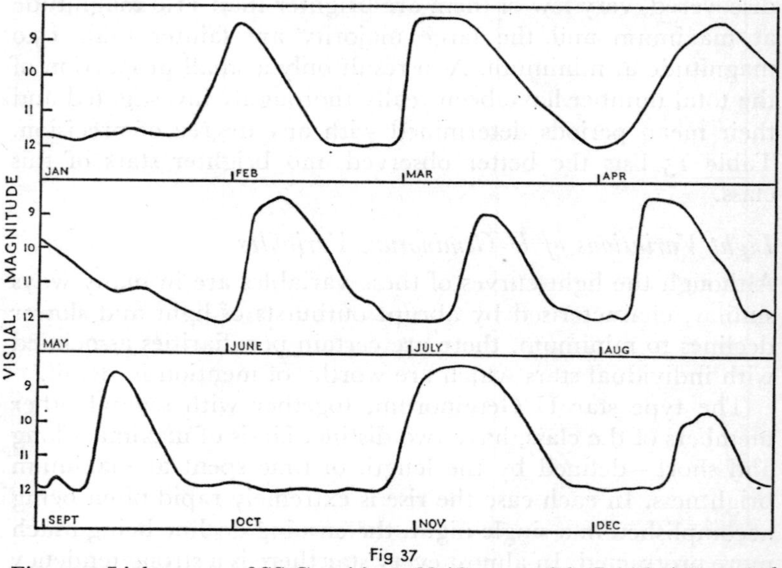

Fig. 37 – Light curve of SS Cygni in 1966 (*Courtesy of the British Astronomical Association*).

from 1 to 10 which is a measure of the time spent above tenth magnitude. Because of irregularities in the curve, even around maximum, it is easier to define the dates when the star is at magnitude 10.0 as it brightens and fades, than when it reaches its maximum brightness. Of the total maxima observed to the end of 1966, by far the greater number have been of type A. As with U Geminorum, long and short maxima show a tendency to alternate although this is not an invariable feature.

Another of these variables having a peculiar light curve and one which has been extensively followed by the author in recent years is SU Ursae Majoris. This star has perhaps the shortest mean period of all the U Geminorum variables and being circumpolar in northern latitudes can be observed all the year round. Usually it varies between magnitudes 12.2 and 14.5 and remains at maximum for only one or two days before fading rapidly. Because of its general faintness and the very short periods of time spent above minimum it is not an easy

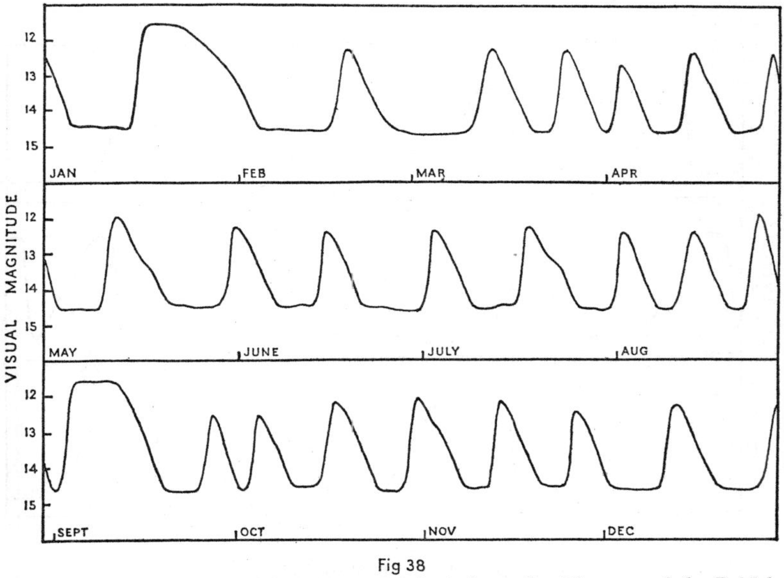

Fig. 38 – Light curve of SU Ursae Majoris in 1962 (*Courtesy of the British Astronomical Association*).

star to observe and undoubtedly many maxima are missed because of bad weather. During the summer months too when it lies below the pole, twilight often hampers its observation and a large aperture is required to observe it at minimum.

About every six months or so however, a 'supermaximum' occurs with the star brightening slowly to just above eleventh magnitude and remaining above minimum for between fifteen and twenty-five days. The general behaviour of this variable is well illustrated in Fig. 38.

We come now to a third member of this class which exhibits

anomalous behaviour. At first sight, RU Pegasi would appear to be quite a normal U Geminorum star. The maxima are well defined and the period varies little from the mean of seventy days. A close inspection of the light curve over the past thirty years or so however reveals two curious facts. The extreme minimum magnitude is given as 13.1 whereas it has seldom, if ever, been seen fainter than 12.7 magnitude. Whether this represents a real change in its magnitude at minimum is not

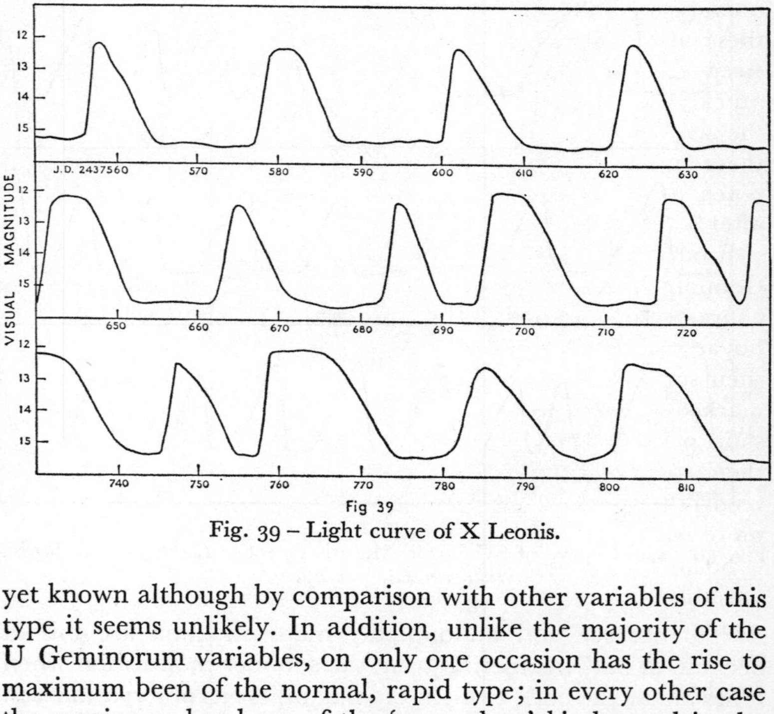

Fig. 39 – Light curve of X Leonis.

yet known although by comparison with other variables of this type it seems unlikely. In addition, unlike the majority of the U Geminorum variables, on only one occasion has the rise to maximum been of the normal, rapid type; in every other case the maximum has been of the 'anomalous' kind noted in the light curve of SS Cygni (Fig. 37). The reason for this odd behaviour is not yet fully understood especially as in other ways, e.g. spectral properties and the radial velocity curve, it so closely resembles the other members of the class.

A further variable which has, on one occasion, shown a peculiarity in its light variations is X Leonis. This is one of the short-period U Geminorum stars, the maxima usually occurring close to the mean period of twenty-two days but for a time in 1961 the maxima appeared grouped in pairs (Fig. 39). After several such cycles, the star returned once more to its more usual behaviour.

Lastly we should mention VW Hydri which has been well observed by southern astronomers being almost as bright as SS Cygni at maximum although the minimum is about one magnitude fainter. As with U Geminorum, two kinds of maxima have been observed in the light curve of this star but these are markedly different in shape. In the first there is a steep rise to a very sharp peak, the variable fading quite quickly to minimum. The second type – a flat maximum is, on the average, about a magnitude brighter. Normally, during these far less common outbursts, the star takes several days to reach its maximum brightness and fades slowly to minimum after a series of minor fluctuations.

From the above discussion it is clear that individual U Geminorum variables differ quite appreciably in their light characteristics although they all share certain features of the novae such as the general abruptness of their brightening and their subsequent decline. At the same time however they differ markedly in their much smaller amplitudes, their comparatively short periods (which do however cover quite a wide range), and their spectra at maximum. When we come to consider the recurrent novae it will be seen that here the similarity is even more noticeable.

So far we have discussed only the maxima of these remarkable variables. The minima which, in general, require large instruments for their detailed observation, have been much less studied; this is only to be expected since only a handful are brighter than magnitude 15.0 at minimum. The behaviour of these stars at this phase is, however, of far greater importance in a study of their physical nature than that at maximum. Those which are bright enough to have been examined photoelectrically show rapid variations in their light with a time scale of between one and ten minutes and a range of a few hundredths of a magnitude. These fluctuations are very like those found in the old novae and other nova-like objects and taken in conjunction with the spectral variations have played a leading role in the elucidation of the physical characteristics of these stars. It is to this subject that we must now turn.

The Nature of the U Geminorum Variables

One of the major problems which has faced astronomers in their efforts to learn something of the nature of these peculiar stars is the difficulty of obtaining accurate distances from which estimates of their intrinsic luminosities may be determined. Early attempts utilised the trigometric method of parallax and were found to yield discordant results. The determinations for SS Cygni, for instance, varied from −0.012 second of arc indicating an extremely remote, very luminous object, to that found by Kukarkin and Parenago of +0.038 from the star's proper motion. The latter value gives us an absolute magnitude at minimum of +10 and would imply that SS Cygni is a white dwarf star. The colour index at minimum, that is the difference between the visual and photographic magnitudes, is +0.62 according to Walker, rather like Nova Herculis (1934) at the present time.

The distances of some other U Geminorum stars has been estimated by Smart and Herbig, the mean value found being only about 200 light years. From this we would expect very little interstellar reddening and since, at maximum, the spectra of U Geminorum and SS Cygni are very nearly continuous, we may make the assumption that these stars radiate as black bodies at maximum light yielding temperatures for these two variables of 12,000 and 15,000°K respectively. This is consistent with an absolute magnitude of +9.5 as found from some of the early determinations of parallax. However, this is not in agreement with stars of spectral classes G5 to K0 and a mass similar to that of the sun which, as we shall see later, have been calculated for the secondary components of these systems. We are therefore reluctantly forced to the conclusion that for some reason these stars are far less luminous than would normally be the case.

There is however, one other class of variable which shows this peculiarity. These are the W Ursae Majoris stars which have been fully described in an earlier chapter. There is, too, a further point of similarity between these two classes of variable. The small peculiar velocities of the U Geminorum stars indicate that they belong to the galactic disc and not to the galactic halo and accordingly it has been suggested that they are a later stage in the evolution of the W Ursae Majoris variables.

We have already seen that in the W Ursae Majoris system, the primary star loses mass in the form of hydrogen which may be either lost into space or trapped by the secondary component. We can postulate therefore that as this mass loss from the primary continues, the star's evolution is speeded up to the point where it becomes a white dwarf and once this occurs, the stage is set for a reversal of the gas flow which now moves from the secondary back to the primary, a condition which has been recently found to prevail in some of the U Geminorum stars.

This theory is extremely attractive and agrees with observation on many points. But it still presents us with unanswered questions. What is the mechanism by which a star can lose as much as a quarter of its mass and become a white dwarf in the process? We do not know. Nor do we know the relationship between the binary condition and the nova-like outbursts of the primary. Furthermore, we would expect to find some variable stars of an intermediate nature which would provide us with an insight into the apparent fact that the violent outbursts occur only in those binary systems where the masses of the components are very nearly equal.

Our discussion so far has been based upon the assumption that these variables are binary systems. We must now examine the evidence for this which has only been put forward within the last decade or so.

Spectra of the U Geminorum Stars

It has already been said that SS Cygni is visually the most completely observed of these variables owing to its comparative brightness at all phases. This is true also of its spectroscopic variations since its magnitude of 12.1 at extreme minimum makes it a relatively easy object for such investigations. When faint the star is yellowish in colour becoming gradually bluer as it brightens, the change being due to a strengthening of the underlying continuum with the peak intensity shifting towards the violet. Shortly after minimum, bright emission lines appear which become progressively weaker as the star rises towards maximum with a corresponding increase in the intensity of the hydrogen absorption lines, double absorption lines making their appearance in the spectrum for a short period, while at maximum the spectrum is dominated by a broad absorption.

When passing through a maximum of Campbell's type A, that is a broad maximum, the variable is of spectral type A1 corresponding to a surface temperature of 12,000°K but at minimum the spectral type is dG5 with a temperature somewhat less than 5,000°K. This extremely wide variation in surface temperature is almost impossible to explain on the basis of a single star, especially in view of the comparatively short mean period.

In 1956 it was announced by Joy that SS Cygni is a spectroscopic binary with a period of only 6 hours and 38 minutes, the components being of spectral types sdBe and dG5. Around the same time, Walker showed that Nova DQ Herculis (1934) is also an eclipsing binary with a period of 4 hours and 39 minutes and it was perhaps inevitable that these results should lead to the speculation that all of the U Geminorum stars are close binary systems.

Investigations carried out by Kraft have now proved that in addition to SS Cygni; U Geminorum, SS Aurigae and RU Pegasi are all binaries with periods of less than nine hours and that the spectrum of EY Cygni is composite. Since the orbital planes of these variables, like those of the Algol eclipsing binaries, will be inclined randomly to our line of sight, it would be too much to expect all of them to show evidence of eclipse. The fact that such a large proportion of those investigated are binaries is circumstantial evidence that all are physically connected doubles.

In those cases where it is possible to detect a late type spectrum, it is found to be either dG or dK. The ratio of the masses of the two components is almost unity with the yellow star slightly more massive than the blue. In every U Geminorum variable so far proved to have a binary nature, the orbit as derived from the radial velocity curve is approximately circular. (Fig. 40.)

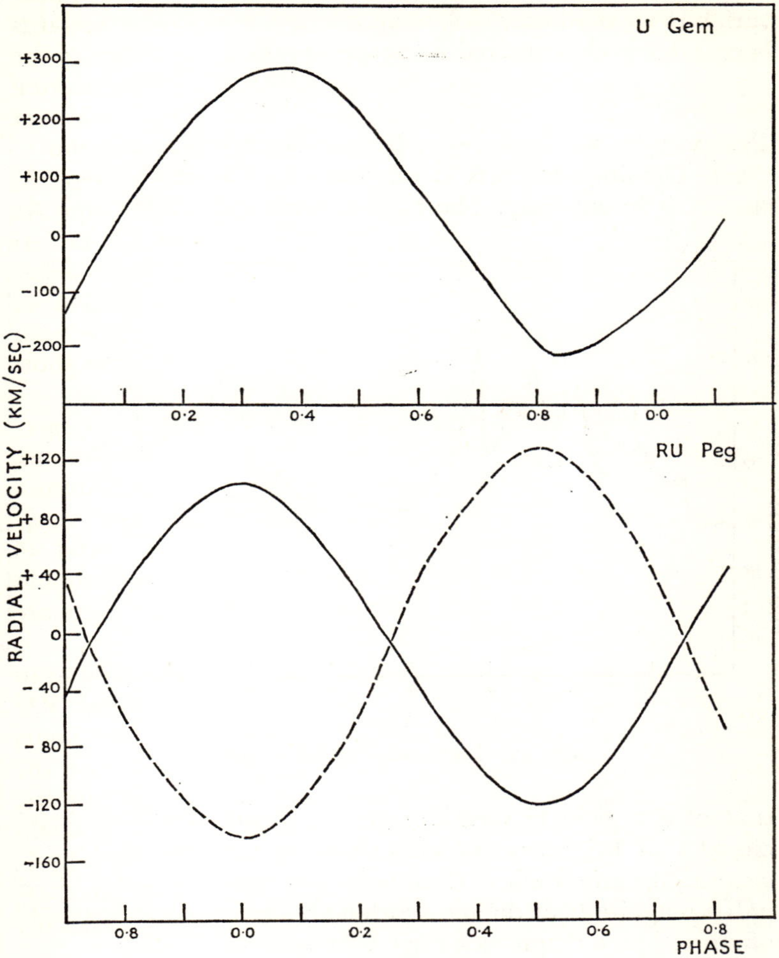

Fig 40

Fig. 40 – Radial velocity curves of two U Geminorum variables (*Astrophysical Journal by kind permission of the University of Chicago Press*).

Calculations of the masses of the components indicates that the late type stars are all roughly as massive as the sun and since the blue, nova-type companions are just slightly less massive, this places them in the white dwarf class.

To obtain further information on the nature of these peculiar variables we shall now look at an extremely odd star which has been extensively observed for many decades.

AE Aquarii

This peculiar star is not, strictly speaking, a typical member of the U Geminorum class at all. Indeed, it is very difficult to classify it in any way. The light curve is quite different from

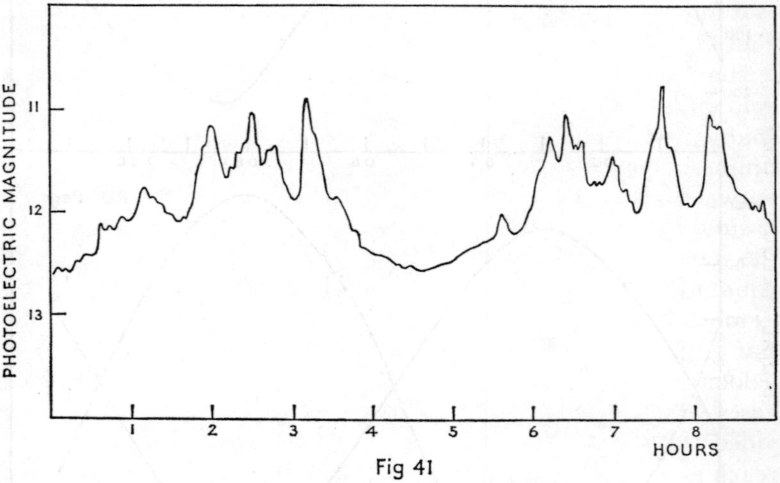

Fig. 41 – Light curve of AE Aquarii.

that of a typical U Geminorum star being subject to rapid changes of brightness of an extremely complex nature. In general, the amplitude of these nova-like eruptions is only about half a magnitude at the most although a few isolated outbursts of more than a magnitude have been suspected. There is little, if any, periodicity at all in these irregular increases in light although they appear to have a cycle of about an hour. The general appearance of the light curve (Fig. 41), is somewhat similar to that of SS Cygni at minimum but on a greatly magnified scale. Spectroscopically, the star does bear some resemblance to the U Geminorum variables and we know it to be a binary system with a period of almost seventeen hours.

Like the present class of variables it too consists of a very hot B type star and a cool dwarf companion of spectral type K5.

The importance of this particular variable in the present discussion is that it provides us with a model on which to base our interpretation of the stars of the U Geminorum class. Since the K type star has a mass at least as high as the sun its stay on the main sequence must necessarily be quite short as its evolution will be much faster than that for similar, but less massive, K type stars. Consequently it will now be expanding as it moves to the right of the main sequence. The gravitational field of the system will however limit this expansion.

Here it is necessary to go into the effect of a gravitational field on the expansion of a star in a little more detail as this will enable us to understand the dynamics of such systems. In the simplest case of a single star, its gravitational field will extend outward equally in all directions, that is in the form of a sphere around the star. At some distance from the centre of gravity of the star a point will be found where the pull of gravity will be equal to the energy of motion of any atom (or molecule) which is moving away from the centre of gravity by reason of the star's expansion. This outer surface is known as the zero-velocity or equipotential surface and we can see that all of the gases which make up the star will remain held inside it provided that their velocities diminish to zero by the time they reach it. In other words, they will remain permanently trapped inside this spherical surface since their velocities of outward motion are not sufficient to enable them to break free of the gravitational pull of the star.

Now what is the situation when we come to examine the more complex case of a close binary system? Here we have also to take into account the interfering gravitational field of the other companion and we find that this has the effect of distorting our previous spherical model into two lobes, rather like a dumb-bell, one lobe surrounding each star and meeting at a point which is known as the inner Lagrangian point.

On this basis we may put forward the following picture of a close binary system such as AE Aquarii and possibly for all of the U Geminorum variables. The more massive component, in this case the yellow-orange K star, having left the main sequence has expanded until the material of which it is composed has completely filled the volume within its own lobe whereas this is not the case for the B type companion which is probably still in the process of evolving towards the main sequence and as a result is not expanding. With this picture as a starting point it is easy to see that although the K star is restrained from any further expansion, there is still the urge for it to do so as it continues to evolve. The only way in which this can now happen is for matter to escape from the now completely filled lobe and this may only occur if some additional energy can be imparted to the gas. This is what actually happens. The important question is where will it occur?

The most likely place is through the inner Lagrangian point for it is here that the opposing gravitational pull of the B star is strongest and consequently a stream of gas from the K component will escape through this point into the unfilled lobe surrounding the blue companion. We have seen earlier in this book that this is the case for the Algol variables where there is very strong observational evidence for the existence of just such a moving gas stream within these systems. Let us now see if there is similar evidence for this in the case of AE Aquarii.

Unfortunately it is not easy to obtain suitable spectra of this variable since it is quite faint and the work is further hampered by the infrequent, irregular outbursts which complicate the spectrum but sharp, narrow lines of ionised calcium have been found and when coupled with the radial velocity curve for the K type companion show that they are associated with the cool star as it moves in its orbit. Further indirect evidence is provided by a study of the blue star itself since it is found to be far brighter than a normal white dwarf and a possible explanation of this anomaly is that gravitational infall of matter from the yellow-orange component contributes to its increased brilliance.

It is of special interest here to consider how this binary system must have evolved to reach its present condition. Earlier, when describing the Algol variables we saw that the evolutionary process could only be reconciled with observation if we assumed that there is interchange of mass between the two stars resulting in the rapid evolution of the initially brighter component away from the main sequence so that it eventually became the fainter of the two.

A similar evolutionary picture may be postulated for AE Aquarii, namely that when originally formed it was the blue star which was the more massive of the two and this being so, it would be the first to begin the process of expansion, losing its outer layers and therefore much of its hydrogen. Once a certain stage of evolution had been reached this star would then commence to contract towards the white dwarf stage as we find it now. In the meantime, the orange-yellow companion will have begun to expand and having filled its lobe of the equipotential surface is now losing mass through the inner Lagrangian point thereby reversing the original process of mass loss.

Reverting to the typical U Geminorum variables therefore we can visualise them as consisting of a hot B type dwarf and a far larger, cooler companion. One significant difference between the main class of variables and AE Aquarii has been found from a detailed study of the spectrum of the latter particularly during the changes which take place at the time of the rapid outbursts. As the star brightens, the emission lines of ionised calcium and of hydrogen increase in strength and broaden perceptibly while at the same time a bright emission appears to swamp the narrow absorption lines due to the late type star. There is, however, no consistency about the broadening of the emission lines. Sometimes they are found to be highly asymmetrical, broadening preferentially either towards the red or the violet end of the spectrum; at other times they may even broaden in both directions simultaneously. All of this evidence thus points to the fact that unlike the outbursts which occur in the U Geminorum stars, those in AE Aquarii are far more localised in nature and may even occur at different parts of the system. Since, however, we know very little concerning the true nature of these outbursts we cannot say whether there is any tendency for them to happen at one particular point or not. At the moment we must await the results of further investigations which will almost certainly throw more light on the physical and dynamical properties of this very intriguing variable.

Returning to the Geminorum variables as a whole, the major problem which as yet remains unsolved is the nature of the outbursts of these stars. In the case of the novae, especially the recurrent novae whose light curves are similar to those of the U Geminorum stars (although on a more expanded time scale), the problem is greatly simplified since their spectra show evidence for the actual ejection of mass. For the U Geminorum stars there is no spectroscopic data whatever for an expanding envelope around these stars during an outburst. We are therefore forced to seek some other explanation for their abrupt increases in brightness.

It appears likely, although there is no absolute proof of this, that the blue star, or some region in its proximity, is responsible for these outbursts. Quite clearly a simple model will not explain the behaviour of these variables, the observational data available at the moment only serves to illustrate the extreme complexity of such peculiar systems. It is possible, however, to distinguish three probable origins of the sudden outbursts, namely the blue star itself, the ring or disc of gas surrounding this star, or a very diffuse gas which fills the equipotential lobe around the hot Be star. If we can apply the model of AE Aquarii to these variables as a whole, then the equipotential surface has a diameter approximately equal to that of the sun.

It is unlikely that the blue star itself is the seat of these eruptions since the magnitude of the light changes is quite appreciable (between four and five magnitudes) and if this were simply a case of mass being ejected, either as a spherical shell or as a streamer thrown out in one preferred direction, we would expect the spectrum to reveal this.

We are on even more uncertain ground when we consider the surrounding shell as a possible cause of the rapid increases in brightness. Unlike the Algol variables we cannot explain them on the basis of eclipse by the secondary companion. There are several arguments in favour of a total eclipse of the blue companions, although almost certainly not of the gaseous ring, by the yellow components but at most the depth of primary minimum due to eclipse amounts to one magnitude. There is also no mechanism known at present by which the material in this rotating ring may suddenly increase in temperature and luminosity.

We finally come to a suggestion by Wallerstein that the inner

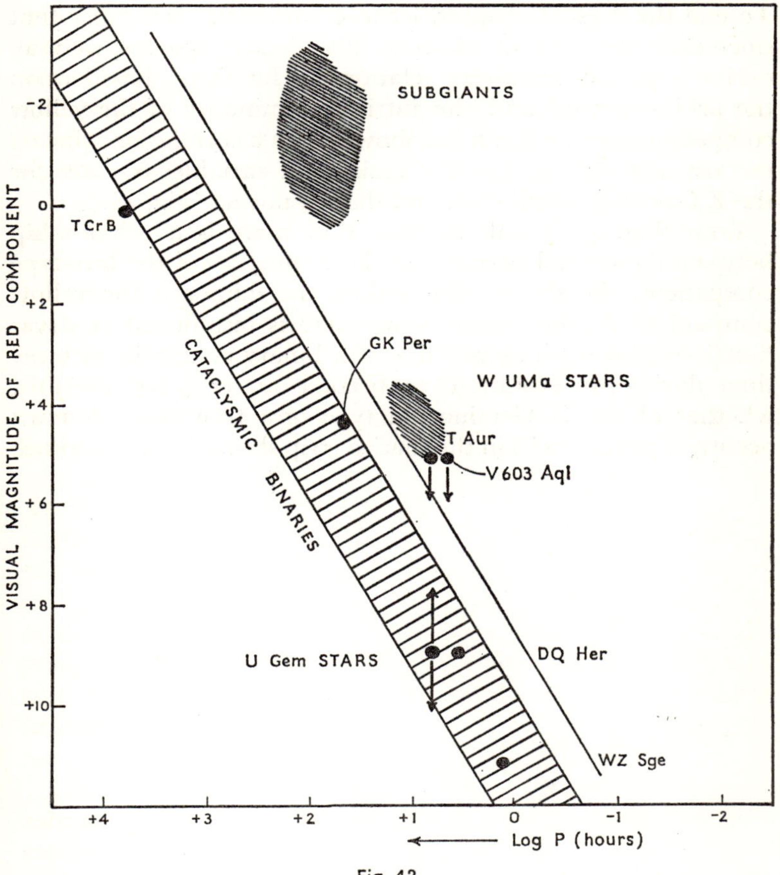

Fig. 42 – Period-luminosity relationship of cataclysmic variables (*Astrophysical Journal by kind permission of the University of Chicago Press*).

Lagrangian surface surrounding the hot star is filled with a diffuse gas which is invisible when the variable is at minimum brightness but which becomes optically thick at the time of a typical outburst. Why the opacity of this gas should change in such an abrupt manner is not yet known although electron scattering has been put forward as a possible explanation.

Period-Luminosity Relationship for U Geminorum Variables

To end the present chapter, let us consider the possibility that since these stars are, in all probability, binary systems, we may derive a period-luminosity relationship for them, based upon the orbital period and the intrinsic luminosity of the yellow component star. As Kraft has shown, such a correlation appears to exist, not only for the U Geminorum variables but also for the Z Camelopardalis stars and the old novae. (Fig. 42.)

From Fig. 42 it will be seen that there is a relationship between the orbital period and the luminosity of the late-type component, the shorter the period, the fainter is the yellow companion. At the present time there is insufficient observational evidence to indicate that if a binary system is, at some time during its evolution, to become an eruptive variable, whether of the U Geminorum type or a true nova, it must occupy a position within the shaded strip shown in the diagram.

15. Z Camelopardalis Variables

The Z Camelopardalis variables constitute one of the smallest of all the classes of variable stars. For many years they were included with the U Geminorum stars discussed in the previous chapter and it is only comparatively recently that they have

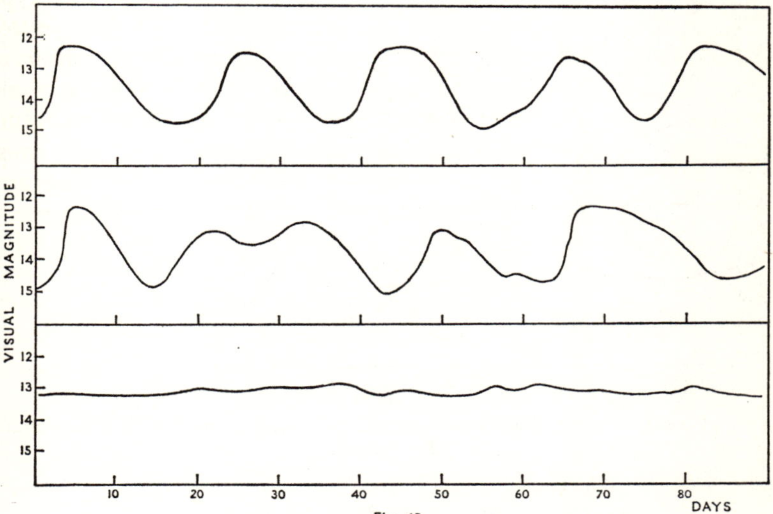

Fig. 43 – Light curve of TZ Persei (*Courtesy of the British Astronomical Association*).

been placed in a class of their own. In their general features they are very like the variables of the preceding group. They are subject to the same unpredictable, abrupt nova-like outbursts and often they show alternate long and short maxima. (Fig. 43.)

When we examine their light curves more closely we find that in general their amplitudes are smaller and their mean periods shorter than the U Geminorum variables although there is considerable overlap, especially as far as the periods are concerned. But the main distinguishing feature of the Z Camelopardalis variables, and one which was not recognised for some time after several had been discovered, are the 'standstills' during which the star will remain at almost constant brightness for an indefinite period, sometimes for only a few days but on occasions for as long as a year or eighteen months

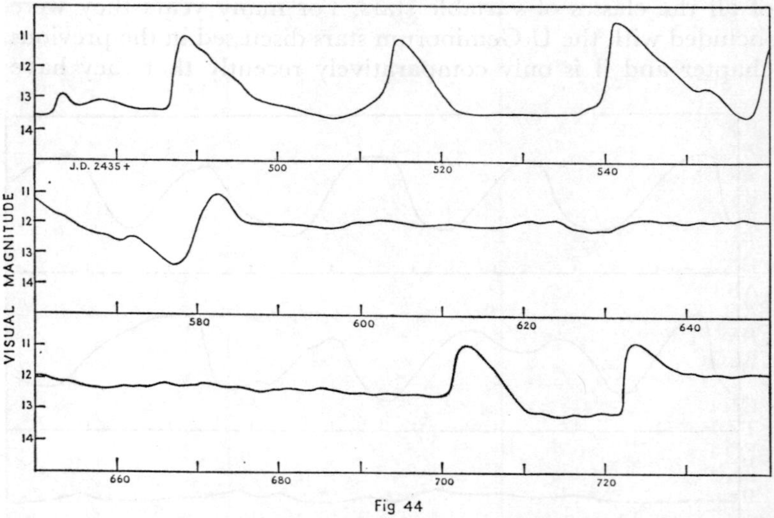

Fig. 44 – Light curve of Z Camelopardalis (*Courtesy of the British Astronomical Association*).

as happened with TZ Persei during 1962.

The 'standstills' almost always occur about a third of the way from maximum to minimum and always on the descending branch of the light curve with only slight fluctuations of brightness amounting to between 0.3 and 0.5 magnitude. There have also been periods when these stars behave in a completely irregular manner as was well observed with Z Camelopardalis in 1956–57. Almost inevitably a 'standstill' is terminated by a continued fall to minimum after which the more normal behaviour is resumed but on at least one occasion Z Camelopardalis actually rose to a maximum following a protracted period of almost constant brightness. (Fig. 44.)

The U Geminorum variables as we have seen are, in the main, very faint stars and the same applies to the Z Camelopardalis sub-group with the result that they are difficult to observe without the use of large instruments. Perhaps it is for this reason that so few are known. At the present time only about

Table 16

Z CAMELOPARDALIS VARIABLES

Star	Maximum	Minimum	Period** (days)	Spectrum
RX And	10.3	13.6	14.1	Pec
Z Cam	10.2	13.4	20	Pec
SV CMi?	13.0*	16.3	—	—
BS Cep	14.0*	15.9	40	—
BP CrA	13.9*	15.6	13.5	—
HN Cyg?	13.8*	15.9	—	—
AB Dra	12.0*	15.8	12	Pec
AH Her	10.6	13.9	19.6	Pec
KW Her?	13.8*	16.5	—	—
BI Ori	13.2	16.6	24.6	—
CN Ori	11.8	14.7	19.2	Pec
TZ Per	12.1	15.4	17	Pec
FO Per	13.8*	16.2	11.3	—
BX Pup	13.8*	15.8	18	—
V735 Sgr	13.5*	16.5	25	—

* Photographic magnitude
** Mean period

fifteen variables of this class have been discovered and observed sufficiently for a mean period to be assigned to them and since with some of these stars the number of 'standstills' are relatively few the inclusion of some of the fainter members in this class must be regarded as tentative. A great need at the present time is for long and comprehensive observation of this interesting class of variables using the largest apertures available for such work.

Table 16 presents the available data for these stars. For many, as will be seen, spectroscopic data is missing but there is no reason to believe that the spectra of these faint members differ markedly from the others.

Like the U Geminorum variables, these stars exhibit their own peculiarities in their light curves but since irregularities occur frequently in the behaviour of these variables they will not be mentioned in detail here. One anomaly however is worthy of note. AH Herculis has been observed continuously by the author for the past eight years using a 13-inch reflector and on no occasion has a minimum been observed in which the star was fainter than 12.8 magnitude visual, but from three photographs taken when the star was visually estimated at 12.7, the photographic magnitude was found to be about 13.8. This yields the very high colour index for this star at minimum of +1.1 magnitude similar to that of a late K or early M type star. This is not in good agreement with the generally accepted view that the secondary components of these systems are yellow dwarfs similar to the sun.

Spectra of the Z Camelopardalis Variables

Not only are few members of this class known but even fewer are brighter than twelfth magnitude at maximum while at minimum the majority are very faint objects. Our contemporary knowledge of the spectra of these stars is therefore based on a very small number indeed. Like the U Geminorum variables which have been more thoroughly investigated, these stars have nearly continuous spectra at maximum light, the energy distribution being essentially similar to that of a typical A or B type star. At times, lines of hydrogen are visible but these are always broad and faint making identification difficult.

One interesting feature which has arisen from the study of their spectra at maximum is that here too there is no sign of any forbidden lines and they do not appear to be surrounded by an expanding shell of hydrogen as is found in the case of the novae. It appears quite likely that the mechanism suggested for the U Geminorum stars is also responsible for the outbursts of these variables. It would be wrong to be dogmatic on this point since the 'standstills' add a further complication in this case and far more research is necessary before we know anything definite about the nature of these outbursts.

The most important result to come out of recent work on the dwarf novae has been the discovery, mentioned in the previous chapter, that most, if not all, of these variables are binary systems. This is true of the Z Camelopardalis stars too and explains many of the peculiarities in their spectra. At maximum light, the blue component is clearly dominant while for most of the light cycle, and particularly around minimum, the G type star dominates the spectrum producing the absorption features with central emission lines as the star fades and finally the double absorption lines at minimum.

Radial Velocities of Z Camelopardalis Variables

The problem of measuring radial velocities for the two components of a Z Camelopardalis variable is even more acute than that for the stars of the U Geminorum class since not only are fewer of these stars known but only two or three are sufficiently bright for good spectra to be obtained at high dispersion. However, from a study of RX Andromedae and Z Camelopardalis itself, radial velocity curves have been obtained (Fig. 45) and for both of these stars the curves are highly asymmetrical showing that the components move, not in circular orbits as has been found for the related U Geminorum variables, but in elliptical orbits with a high eccentricity.

This at once raises an extremely important question. Are the components of all the Z Camelopardalis systems revolving in elliptical orbits and if so, is this one of the fundamental differences between the two classes which might possibly give us an explanation of the curious 'standstills' for these stars? It should be said at once that we know very little of the reasons for these periods of comparative inactivity and any theory is bound to be mere speculation.

It may however be of interest to speculate on one possible hypothesis. Let us suppose that all of these variables do possess elliptical orbits. Let us furthermore assume that it is not a change in the blue star but one in the yellow companion which is directly responsible for the 'standstills'. Do these suppositions bring us any closer to solving the problem?

One direct result of the high eccentricity of the orbit is that at times the two stars will be much closer together than at others and if an outburst of the blue component (for the moment we will not go into the form which such an outburst may take) occurs at the moment when they have approached within a certain critical distance, this may be sufficient to produce a corresponding effect upon the yellow, non-variable component. If this should happen we can visualise a situation where high temperature gases and radiation from the vicinity of the blue star may then ionise the surface layers of the companion, tem-

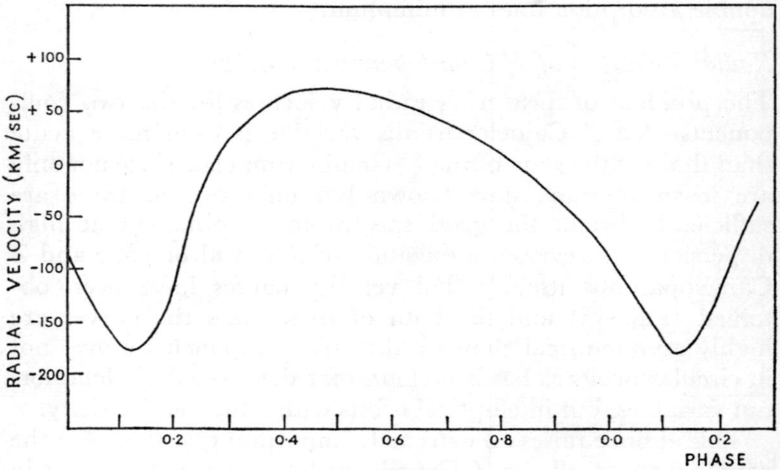

Fig 45

Fig. 45 – Radial velocity curve of RX Andromedae (*Astrophysical Journal by kind permission of the University of Chicago Press*).

porarily raising the temperature and increasing the total radiation, rather like the Reflection Effect exhibited by many of the eclipsing binaries but on a far larger scale.

On this hypothesis the 'standstills' would therefore be due, not to the hot, blue variable star, but to changes induced in the atmosphere of the yellow component. Some confirmation for this idea is provided by the observational fact that virtually all 'standstills' so far noted begin about a third of the way down from maximum to minimum light, suggesting that they are somehow initiated by the primary outbursts, the short time lag between maximum and the onset of a 'standstill' being that required for the luminosity of the faint yellow star to rise to its eventual temporary maximum. (Fig. 46.) We would not expect the U Geminorum stars to exhibit such periods of relative inactivity since revolving in circular orbits as they do, the stars will not approach within this critical distance. Some confirmation of this idea may possibly be provided by an examination of the spectra of these stars under high dispersion at the time of a 'standstill' when changes in the spectrum of the late type star might be expected.

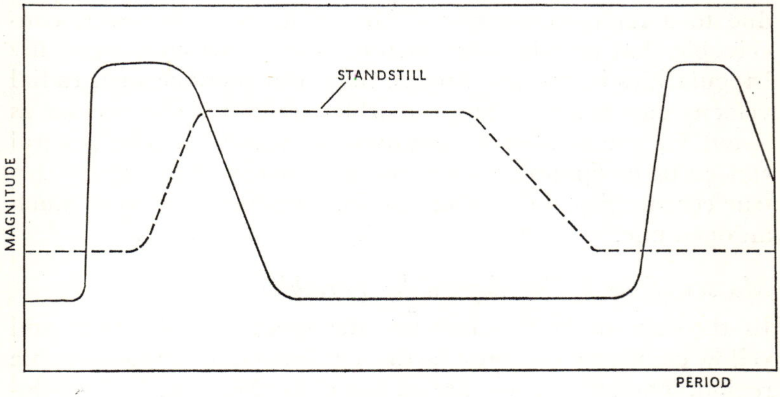

Fig 46

Fig. 46 – Suggested light curves of the components of a typical Z Camelopardalis variable.

There are, of course, several criticisms which this hypothesis must face. For example the duration of individual 'standstills' vary tremendously, some lasting for only a few days, others for many months. It may be significant however that the longest 'standstills' recorded for both RX Andromedae and TZ Persei followed very closely upon the brightest maxima ever observed for these two stars. We must also take into consideration the fact that the orbital periods of these stars are very short even compared with the time spent at maximum and statistically we would expect more 'standstills' than are actually observed.

We are assuming, however, that the outburst of the primary takes the form of a spherically expanding shell and as we have seen the present spectroscopic evidence is all against this view.

There remains the possibility that in these variables the material is ejected preferentially in one particular direction, more in the form of a large jet than an envelope of gas. If we accept this argument then the relative infrequency of 'standstills' may be partially explained at least on the assumption that none will occur unless the ejection of gas in the right direction, e.g. towards the yellow dwarf companion, coincides with the moment of closest approach.

All of what has just been said, of course, is based upon the assumption that the asymmetry of the radial velocity curves is due to a high eccentricity of the orbit. It is, however, conceivable that certain other factors in these systems, especially irregularities in the gas stream, may also produce such radial velocity curves and in fact the orbits are essentially circular as found for the related U Geminorum variables. The general concensus of opinion is that the asymmetry does represent a true eccentricity of the orbit but the question is still very much an open one.

Masses of the Z Camelopardalis Variables

In the case of U Geminorum, the spectra of the blue and yellow components cannot be distinguished in the photographic region. Fortunately we are able to do this with Z Camelopardalis and can state with some confidence that the mass of the blue primary is 0.57 times that of the sun, the corresponding figure for the yellow companion being 0.68 solar mass. The stars which make up this system are therefore somewhat more massive than those of, say, the old nova DQ Herculis (1934) and it would appear that the blue component of Z Camelopardalis is well within the theoretical mass limit for a typical white dwarf.

16. Nebular Variables

So far in our consideration of intrinsically variable stars we have been discussing those whose light variations are due more to internal changes than to any external agency. In this chapter we include a wide diversity of variables which have only two features in common; they are all of fairly low intrinsic luminosity and they show irregular patterns of behaviour in their light curves which are due, in part at least, to the presence of associated nebulosity. Many hundreds of these variables have been discovered in the Orion Nebula and other vast associations of such stars occur in the neighbouring constellations of Auriga and Taurus. In the southern hemisphere, similar huge agglomerations are found in the bright and dark nebulae situated in Corona Austrinus, Carina and Monoceros.

The stars we shall be concerned with here form a very heterogeneous class but in essence they have light curves in which there is very little regularity; some such as the subgroup typified by RW Aurigae have quite large amplitudes while at the other end of the scale those whose prototype is T Tauri have amplitudes of the order of only half a magnitude.

Since their light changes are so erratic, it would appear at first sight virtually impossible to divide them into any individual classes based upon either light or spectral characteristics. Wenzel however has shown that from an examination of their light curves they may be conveniently placed into three groups depending mainly upon the length of time spent above minimum.

GROUP A: These are stars which correspond in the main to the earlier classification of RW Aurigae variables and spend approximately equal times around minimum as at maximum. As we shall see later, a small number of these variables are binaries in which one of the components is a T Tauri star.

GROUP B: Variables which are around maximum brightness for most of the time but are subject to rather abrupt minima, in certain instances this feature is so marked that the light curve is difficult to distinguish from that of a typical Algol variable. In this group are placed the T Orionis variables which are preponderantly A type stars always associated with nebulosity.

GROUP C: This group is composed mainly of the T Tauri

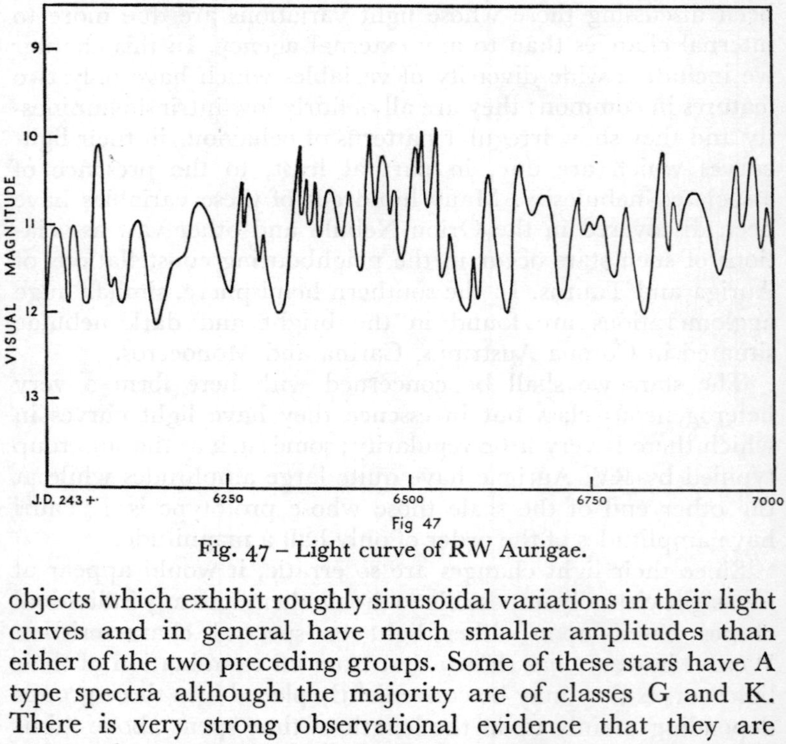

Fig. 47 – Light curve of RW Aurigae.

objects which exhibit roughly sinusoidal variations in their light curves and in general have much smaller amplitudes than either of the two preceding groups. Some of these stars have A type spectra although the majority are of classes G and K. There is very strong observational evidence that they are among the youngest of all the stars and it is very likely that many of them are actually stars in the process of condensing out of the interstellar gas and dust clouds, or stars in which the formation of a planetary system is taking place.

The scheme outlined above provides a very simplified classification of these stars and we must not infer from it that the variables within each group form distinct classes. Overlapping occurs between all three groups and it is also possible to subdivide the first two still further according to the general appearance of the light curve.

Although there is this wide difference between these groups of variables, it is noticeable that examples of every type are found in close proximity, especially in the four great associations – T_1 and T_3 Tauri which lie near the star Chi Tauri and T_1 and T_4 Orionis very close to Lambda Orionis.

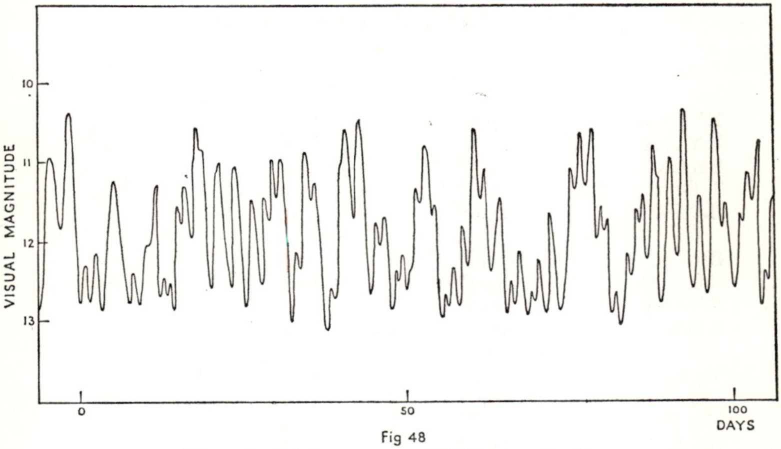

Fig. 48 – Light curve of T Chamaelontis.

RW Aurigae Variables

Visual observations of these curious stars show that there are rapid and completely irregular variations in their light curves, typified by that of RW Aurigae itself. (Fig. 47.) Some stars are known, however, such as T Chamaeleontis where there does appear to be a pseudo-periodic wave superimposed upon the primary light fluctuations. (Fig. 48.)

As a large number of these variables are found associated with the T Orionis stars in the huge gas clouds of Orion, Taurus, Auriga, Monocerotis and Corona Austrinus, there is often the problem, especially with the fainter members, of distinguishing the two types particularly when only their light curves are available for study. In the dubious cases, where it is possible to obtain a spectrum of the variable in question, this often provides a clue since the RW Aurigae stars are almost all G type dwarfs whereas the latter are much earlier in spectral type being mainly A stars.

We find very few of these stars which are brighter than tenth magnitude at maximum and at minimum most of them are extremely faint objects, making it difficult to obtain spectra

Table 17

RW AURIGAE VARIABLES

Star	Maximum	Minimum	Spectrum
RW Aur	9.0	12.0	dG5
UY Aur	11.6*	14.0	dG5
BE Cas	13.5*	14.4	—
T Cha	10.0*	13.2	dG5
BZ Del	13.6*	14.3	—
BR Her	13.0*	14.0	—
RU Lup	9.3*	13.2	dG5e
RY Lup	9.9*	13.0	—
R Mon	10.0	14.2	G5
V426 Ori	12.2*	13.8	—
CO Ori	10.0	13.0	dK0
DN Per	14.8*	16.2	—
BK Ser	14.3*	16.7	—
RR Tau	10.2	14.2	G1

* Photographic magnitude

at this phase. Table 17 lists the available data for some of the brighter stars of this class.

Four of the RW Aurigae variables are in a different category to the others. There may be more like these but if there are they have not been as extensively studied. These four stars are RW Aurigae, UY Aurigae, UX Tauri and UZ Tauri. What makes these stars so important is that they are all visual doubles and from a study of these binary systems one curious result has emerged. The companions have all been found to be red dwarfs of spectral type M with mean apparent magnitudes between 12.5 and 12.9. The primaries however are dwarf G5 stars very similar to the sun, have apparent visual magnitudes only slightly brighter – about 12.0 magnitude. This small difference of only 0.5 to 0.9 magnitude is far less than would be anticipated from their spectral types. There appear to be no distance factors complicating the issue here. Since both stars are dwarfs and being physically connected are at the same distance from us, we would expect there to be a difference of at least three magnitudes in their apparent brightness.

What has gone wrong? Either a mistake has been made in our observations and the M type dwarf stars are not physical companions to the RW Aurigae stars or there is something very peculiar about these red components. There is now so much observational evidence to show that these variables do form physical doubles that we are forced to look to the second argument for a possible explanation.

We start by examining the probable positions of these stars on the Russell-Hertzsprung diagram since this may provide us with a clue to this serious discrepancy. The G type stars are obviously on the main sequence being like the sun in this respect and have not yet begun their evolution to the right. But what of the red companions? One reason for their being over-luminous for their spectral type would be if they were above the main sequence. The problem which now faces us is: can we find any independent evidence that such red dwarfs do exist above the main sequence?

One place where we might look for such evidence is in the very young clusters which are known to contain fairly large numbers of these red dwarf irregular variables. A study of the late-type stars in NGC 2264 shows quite definitely that all of them are appreciably over-luminous. Either they are red dwarfs which have evolved away from the main sequence or more likely they are evolving onto it. This phenomenon of over-luminosity is therefore not unknown and provides us with the most probable explanation of these peculiar binary systems.

The spectra of the RW Aurigae variables show quite strong emission lines of hydrogen, particularly H α, the intensity of the emission lines varying in step with the light phase, being almost invariably stronger as the magnitude increases.

T Orionis Variables

Since the invention of the telescope the existence of many, bright and irregular patches of nebulosity occupying fairly large volumes of space has been recognised. The first instruments were not powerful enough to enable the early astronomers to determine whether or not all nebulae were of the same kind but as telescopes increased in aperture (and therefore in light gathering power) and particularly once long exposure photographs of the heavens became available, there emerged various hints that these objects could be divided into two classes, those which could be resolved into stars and those which remained completely irresolvable. The final, conclusive proof came in 1864 when Huggins used the spectroscope which he had previously employed with great success for stellar spectroscopy, for the investigation of the nebulae.

The spectra of the resolvable nebulae, examplified by the spiral in Andromeda, were found to consist of many absorption lines on a continuous background such as we might expect to be produced by a vast agglomeration of stars of all types. These are known as the extragalactic nebulae, or simply galaxies, which are isolated stellar systems comparable with our own galaxy and containing many hundreds of millions of stars.

The spectrum of a particular nebula in Draco however was quite different, showing only a few bright emission lines and Huggins recognised immediately that such a spectrum could be produced only by a vast cloud of glowing gas. The brightest and finest example of these gaseous nebulae is undoubtedly that in Orion which is just visible to the naked eye on a dark, moonless night. This nebula is about 1,100 light years distant and the visible portion of it must therefore be about thirty light years in diameter. Long exposure photographs of the region have shown however that, in reality, it extends much further than this and its true diameter is far larger.

Many similar gaseous nebulae are now known and since their spectra are essentially bright line emission types, it is clear that the gas must be self-luminous although much of the luminosity is due to the radiation from the intensely hot stars which are embedded within them. The fact too that several of the lines are those of ionised atoms; hydrogen, helium, oxygen and nitrogen is further proof of the high surface temperatures of the embedded stars, predominantly of types O, B and A with temperatures in excess of 12,000°K. Much of the emitted radiation of these stars is in the high-energy ultra-violet region of the spectrum, invisible to the naked eye, this radiation being converted by a process of fluorescence into visible light.

The importance of these gaseous nebulae in the present discussion is that quite a high proportion of these hot O, B and A stars are variable and almost certainly one of the major

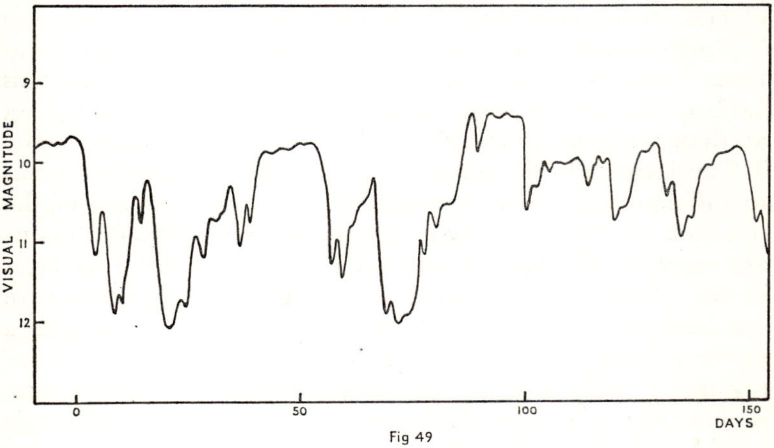

Fig. 49 – Light curve of T Orionis.

causes of their variability is that of changes in the opacity of the gas through which they are passing.

The prototype of this class of variable, of which several thousand are known at the present time, is T Orionis, normally of 9.5 magnitude but fading irregularly, often in the space of only a few days by as much as three magnitudes. These fadings are completely unpredictable (Fig. 49) but since the star appears bright for most of the time it clearly belongs to Wenzel's Group B. This feature of irregular fading is shown in varying degrees by all of the T Orionis variables. In some the irregular minima, although varying in amplitude, are quite well defined, so much so that in one or two extreme cases, e.g.

BO Cephei, the star has been mistakenly classified as an Algol variable of a peculiar kind.

Again, we find that the large majority of these variables are extremely faint. Only a handful are as bright as eleventh magnitude at maximum and since light curves and spectra have been determined for only the brighter members of the class it is inevitable that there should be some overlapping with the RW Aurigae stars already mentioned. Undoubtedly, apart from their spectral distribution which may be indicative either of some real difference in their physical nature or that one group has evolved further than the other, they are very similar objects. Sometimes, they have also been confused with the T Tauri variables discussed below as their light curves do at times resemble each other, more especially if the T Orionis variable has only a small amplitude. The close association of all these types of variable also complicates matters but since the T Tauri stars all have spectra of G type dwarfs this often serves to differentiate them. A more detailed understanding of the situation will undoubtedly be got when a schematic evolutionary sequence for these variables is found. As we shall see later in the chapter, the T Tauri variables are stars which are condensing out of the huge gas clouds although showing a preference for the small dark nebulae which abound in these regions. It may well be that the T Orionis stars are quite similar in this respect differing only in their size and their higher surface temperatures.

In such a vast region such as the Orion nebula for example where stars are being formed, by far the greater number will be evolving deep within the gas and this raises the very important question of how we are able to see them at all since such clouds of gas and dust are highly opaque to visible light. An answer to this has been given only in recent years. Let us consider the case where a blue giant star has been formed out of the parent gas cloud. The surface temperature of such a blue giant will be somewhere in the region of 15,000 to 20,000°K resulting in an exceptionally steep temperature gradient within the gas cloud in the vicinity of the condensing star. Now this has the effect of developing very high pressures around the star, the portions of gas heated by the star forcing the much colder regions away.

These violent disturbances within the gas cloud enable us to see deep into them and observe these stars but it is only in certain instances that we are able to do this while the process of star formation is actually going on. In all other cases, the intervening gas and dust is not expelled from the vicinity of

Table 18
T ORIONIS VARIABLES

Star	Maximum	Minimum	Spectrum
R CrA	10.0*	13.6	F5
S CrA	10.8	12.5	G5
T CrA	11.7	13.5	F0
TY CrA	8.7	12.4	B2
DG CrA	13.1*	15.3	—
T Ori	9.5	12.1	A0
RY Ori	10.5*	14.0	A0
RZ Ori	15.2	16.8	—
SW Ori	12.9*	16.0	A0
SX Ori	14.4*	16.2	—
SY Ori	13.9*	16.0	—
TT Ori	13.2*	15.8	—
TV Ori	13.4*	15.6	K0
UX Ori	8.9*	10.6	A2e
UY Ori	11.5*	13.2	—
SV Sgr	13.5*	16.8	—
DU Sgr	13.2*	16.6	—
UZ Sco	13.2*	16.4	—
AT Sco	13.3*	15.2	—
AU Sco	13.8*	14.7	—

* Photographic magnitude

these stars until they have been formed. The Orion Nebula is one such gas cloud where much of the surrounding material has been forced away but with still sufficient remaining to allow of star formation to continue. We may therefore look upon the T Orionis variables as stars which are in the process of formation and in such a situation the highly irregular nature of their light variations will be due to a combination of two processes; unstable fluctuations associated with the initial stages of star formation and changes in the opacity of the surrounding gas and dust as it is being forced away from the neighbourhood of these blue giants.

From the data given in the following Table, it will readily be seen how the T Orionis stars tend to aggregate in certain fairly well defined regions.

T Tauri Stars

We have just seen how the T Orionis stars are blue giants in the process of forming out of the gigantic gas and dust clouds such as the Orion Nebula. Within this region and also in the areas of the large T-associations which are teeming with RW Aurigae and T Orionis variables, we also find quite small nebulosities where star formation also appears to be taking place. Here however we find, not blue giants, but yellow dwarfs whose spectra, apart from other unique features, show strong emission lines of hydrogen.

These stars, the T Tauri variables, are among the most interesting of all classes of variable stars and are becoming increasingly important in the field of astrophysical research since not only are they very young stars possibly evolving very rapidly onto the main sequence but quite probably, as we shall see later on, they provide us with some evidence of the formation of planetary systems. As they are always associated with fairly small nebulosities, we are not faced with the same problem of seeing these stars since they lie either just within, or on the very edge of these tiny gas and dust clouds and are readily visible. Their low intrinsic luminosity however means that usually they are quite faint objects and large apertures are generally necessary to follow their light changes fully.

The faint nebulosities with which many of the T Tauri variables are associated also often vary in their luminosity due in part at least to the varying illumination they receive from the stars which are embedded in them. On occasions, these stars fade to abnormally deep minima and it is noticeable that these small gas clouds also experience similar fadings whenever these deep minima occur. We do find, however, that in most cases the fluctuations in the light of the nebulosities are greater than those of the stars themselves providing an indication that the gas clouds vary independently in brightness. It may, of course, be that small light changes in the star are accompanied by larger variations in infra-red and ultra-violet radiation from these objects which has a pronounced effect on the luminosity of the associated nebulae.

The light curves of these fascinating variables consist usually of slow, sinusoidal waves of quite small amplitude with a general tendency for more radical fluctuations to appear during the phase when the star is bright. (Fig. 50).

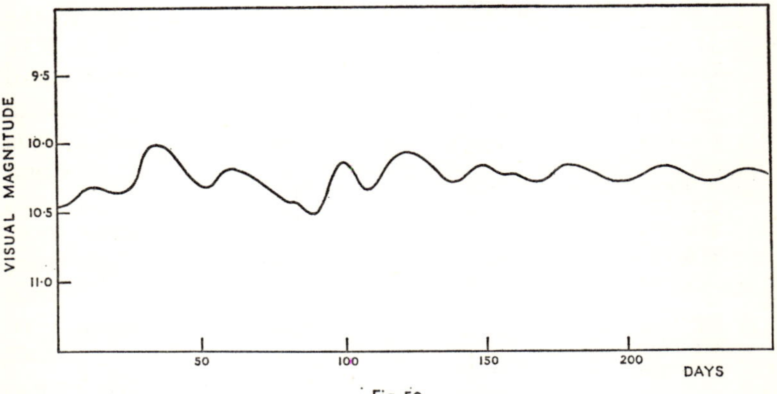

Fig. 50 – Light curve of T Tauri.

A further feature of these stars is that for many of them, although the photographic and visual ranges are almost identical, the visual light curves are much brighter. One of the best observed of this class, RY Tauri, has a photographic range of 10.1 to 12.3 magnitudes whereas the corresponding visual range is 8.6 to 11.0 magnitude (although it has been observed at a deep visual minimum in 1959 when it was as faint as magnitude 12.0).

A list of T Tauri variables is given in Table 19 where it will be seen that unlike the T Orionis stars, all have spectra of type G or later.

Spectra of the T Tauri Variables

The spectra of these stars show several remarkable features. At certain stages of the light variation, a very strong continuous spectrum asserts itself, hiding the underlying late type absorption lines. In some stars, for example DD Tauri, this blue and violet continuum almost completely obscures the absorption spectrum producing a large colour index, making them some of the bluest of this class of variable. This particular star is

Table 19
T TAURI VARIABLES

Star	Maximum	Minimum	Spectrum
SU Aur	9.0	9.6	dG2
GM Aur	13.1*	13.9	dK5e
AN Ori	10.7	11.7	dK1e
GW Ori	10.1*	11.5	dK3e
GX Ori	14.0*	15.3	dK3e
T Tau	9.5	13.0	dG5e
RY Tau	8.6	11.0	dGoe
UX Tau	10.7*	13.4	dG5e
UZ Tau	11.7*	14.9	dG5e
AA Tau	13.7*	15.4	dMoe
BP Tau	10.7*	13.1	dG5e
CI Tau	14.4*	15.2	Ge
CW Tau	13.9*	15.1	dK5e
CX Tau	14.6*	15.5	dMl.5e
CZ Tau	15.7*	17.5	dM2e
DD Tau	14.1*	15.5	dK6e
DE Tau	13.9*	14.6	dM1e
DF Tau	10.7*	13.4	dMoe
DG Tau	11.7*	13.7	Ge
DS Tau	12.1*	13.6	dK4e

* Photographic magnitude

associated with a very small variable nebula designated B10 which completely surrounds the star and also possesses a distinctly blue colour. A further interesting feature of this tiny nebulosity is that the light emitted by it is partially polarised but whether this is due in some way to the enveloped star itself is not yet known.

Two of the brightest of these stars, T and RY Tauri, have been closely studied for many years. The former is associated with Hind's variable nebula NGC 1555, a patch of hazy nebulosity which was quite bright in 1852 when discovered but visible only in large instruments in 1861 when searched for by d'Arrest and completely invisible seven years later. Since that time it has been visible on and off, varying quite appreciably in luminosity and is readily observable in large instruments at the present day and also on long exposure photographs, although it would appear that its general form is not now identical with that which it had when first discovered. The spectrum of the nebula itself shows quite conspicuous emission lines. Since the star itself, being a yellow dwarf does not have a surface temperature sufficiently high to excite the atoms of the gas, we must assume that some other mechanism whose identity is not yet known, is responsible for much of the luminosity of the surrounding mass of gas.

Similarly, RY Tauri illuminates the southern tip of a small, dark nebula discovered by Barnard and designated B214.

The emission lines found in the spectra of those T Tauri variables which have been examined are those of Hβ, neutral oxygen and singly ionised calcium and silicon. The question of the lithium lines which are also visible in these spectra will be discussed separately.

A handful of these variables have been observed with sufficient dispersion to show that the absorption lines are both broad and diffuse leading to the conclusion that their velocities of rotation are high, something of the order of fifty kilometres per second. No other stars near the main sequence and of these spectral types are known with such high rotational velocities and we shall see later that there is some evidence for the preliminary formation of planetary systems among these stars which may account in part for this anomaly.

Since all T Tauri variables so far examined are distinctly brighter than main sequence stars of the same spectral classes, Herbig has put forward the suggestion that these variables are quite young objects, only recently born out of the pre-stellar dust cloud and still undergoing the process of contraction under the influence of gravitational forces. It has further been demonstrated that a star such as RY Tauri which is now a G type dwarf will reach the main sequence in about ten million years and in accordance with theoretical calculations it will become both hotter and smaller in the process, eventually becoming an A7 star. In so doing, conservation of angular momentum will result in a doubling of its rotational velocity to around 100 kilometres per second as is indeed found for main sequence A7 stars.

Similarly those variables which have a later type spectrum, between K and M, will evolve into G0 stars once they reach the main sequence and in their case, they will form more slowly rotating bodies with equatorial velocities very similar to that of the sun. We must remember however that in considering these very cool dwarf stars which are probably just contracting out of a gas and dust cloud, conservation of angular momentum is unlikely to hold during the initial stages of evolution and it is not really surprising therefore that we find anomalous results when dealing with such stars.

Lithium Abundance in T Tauri Stars

Recently a truly remarkable feature of the spectra of these variables has come to light, namely the high over-abundance of lithium in the atmospheres of these stars, between 70 and 100 times that in the sun, comparable to the abundance found on Earth. At present we know very little of the way in which the lighter elements between helium and carbon are formed in stellar interiors and clearly these results from the T Tauri variables are of great importance in providing an understanding of their formation.

Why there should be this large difference in lithium content between these stars and all others is not yet known. Although the survey of T Tauri variables is far from complete, those in which this over-abundance has been found range from the T-associations to the Orion Nebula and it would seem safe to conclude that all of these variables possess a similar lithium content. Either this high proportion of lithium is normal when stars are first formed (here we are, of course, considering only the later Population I stars. It seems very unlikely indeed that any of the older Population II stars will contain so much of this element since they were formed out of the primeval hydrogen of the galaxies), and over many millions of years of evolution the sun has now virtually exhausted its supply or there is some nuclear mechanism taking place within these stars which generates lithium as a by-product.

More recently, infra-red observations of T Tauri and R Monocerotis have been carried out to a limiting wavelength of twenty microns. The results obtained are consistent with a model in which a thick, circumstellar cloud of dust absorbs and then re-radiates the energy which is produced by a central protostar which is still in the process of contraction. This is an important extension of our previous knowledge of these curious objects since if this is so, then the energy released in contraction and much of the angular momentum seem to be gradually being transferred to this envelope of dust and here we have an instance of the first stages in the formation of a planetary system.

The dust cloud surrounding this protostar is almost certainly confined to a disc which is perpendicular to the axis of rotation and not to a spherical shell. In some other T Tauri variables the spectrum does show shell-like absorption lines which are displaced towards the violet indicating an outflow of material from the reversing layer of the star but so far there is no evidence whatever in any of these variables for an inflow of gas as would occur if the star were accreting mass from the enveloping dust.

In connection with these variables which, as we have seen are born in vast numbers within the large gas clouds (here we are including the RW Aurigae and T Orionis variables), some recent observational work by Blaauw and Morgan on dispersing star showers suggests that there have, in the past, been violent cosmic disturbances within certain of these gas clouds. Many star clusters are known which are expanding, indeed it is generally believed that all star clusters must expand once they have been formed. By a careful analysis of these self-disruptive clusters, in particular the ζ Persei cluster which is both close and still reasonably compact, we find that the average stellar velocity of the outward moving stars is twelve kilometres per second. From a further study of associations of stars high on the main sequence, Ambartzumian had previously inferred that this velocity of twelve kilometres per second is at the high end of the scale for such systems and that the majority of really young stars in these expanding clusters have much lower velocities of the order of one to five kilometres per second.

The stars we will consider now have been found by Blaauw and Morgan to possess far higher velocities than this. In particular, three O-type stars, μ Columbae, AE Aurigae and 53 Arietis (the figures obtained for this last star are a little doubtful), are all moving with velocities between 80 and 100 kilometres per second and all three appear to be diverging from a common point which lies somewhere in the region of the Orion Nebula. From their velocities and present distances from this nebula it may be calculated that all three were spatially close neighbours about 2,500,000 years ago. Not only this, but their ages also appear to be of this order of time. Here then we have some evidence that shortly after their formation, these stars were ejected from the region of the Orion Nebula with these very high velocities.

Obviously we cannot explain such high speeds on the basis of an expanding cluster. Is there any other alternative? In the chapter on supernovae we shall see that in certain of the irregular galaxies there is strong observation evidence for super explosions produced by very early Type I supernovae. Now it is not suggested that these three stars were thrown out of the Orion Nebula by an explosion on such a scale as this. For one thing, our galaxy is not one of the irregular ones and so far there is no evidence of such explosions in the spiral galaxies. In addition, the velocities imparted to the gases by these super-explosions are several thousand kilometres per second. But an explosion on a somewhat less violent scale could explain the motions of these stars.

To end this chapter, we must mention one particular variable which is associated with nebulosity but which does not fall into any of the preceding classes. This is the long period variable R Aquarii. The spectrum of this star is highly complex while its light curve shows many deviations from the general form shown by these variables. Long exposure photographs show that R Aquarii is a red supergiant of low surface temperature situated in the midst of a strange nebulosity with which it is intimately connected. There is also an O-type companion to complicate matters still further together with an expanding shell of gas which gives spectral lines very similar to those of the P Cygni stars. Quite clearly, matter is being continually ejected into the binary system, most probably from the O type companion although this is by no means certain.

17. R Coronae Borealis Variables

We now come to a very small and peculiar class of variable, the type star of which, R Coronae Borealis, was discovered by Piggott in 1795. Oddly enough, these stars are the only class of variable for which maximum light is the normal phase. During most of the last century a wide variety of variable stars which could not be satisfactorily fitted into any of the types then known were originally classified as R Coronae Borealis variables, usually on the basis of convenience rather than from any detailed analysis of their light and spectral variations, and as a result this class became one of the most heterogeneous known. Since about 1920 however our knowledge of this unique type has increased enormously and the large majority of these variables have been discarded and at the present time only a score or so of these stars are included in this group.

R Coronae Borealis is normally a star of the sixth magnitude, just visible to the naked eye, in the small but quite conspicuous constellation of the Northern Crown and for months, sometimes as long as nine years, e.g. from 1926 to 1934, it remains at almost constant brightness, any fluctuations not usually exceeding 0.5 magnitude. At irregular intervals however it will suddenly fade without any warning whatsoever, generally to around twelfth magnitude but on occasion it has faded below magnitude 14.0. After passing through these deep minima the star will commence to climb back to its original brightness but in almost every case the ascending branch of the light curve is marked by quite pronounced fluctuations. Indeed, there may be two or more deep minima before the star finally attains sixth magnitude. Recently, in June 1962, R Coronae Borealis commenced to fade and passed through four, possibly five, such minima, being still slightly below normal maximum at the time of writing.

As may be seen from a study of the light curve of this star (Fig. 51), the variations are extremely irregular and unpredictable, there being no way of telling when the star is likely to begin its erratic behaviour.

There has naturally been a great deal of speculation on the basic cause of the peculiar light changes of this class of variable and as yet this is not known with absolute certainty. It is clearly very doubtful if these irregular fadings can be attributed to any mechanism similar to that put forward to explain the periodic light changes of the Cepheids and the long period

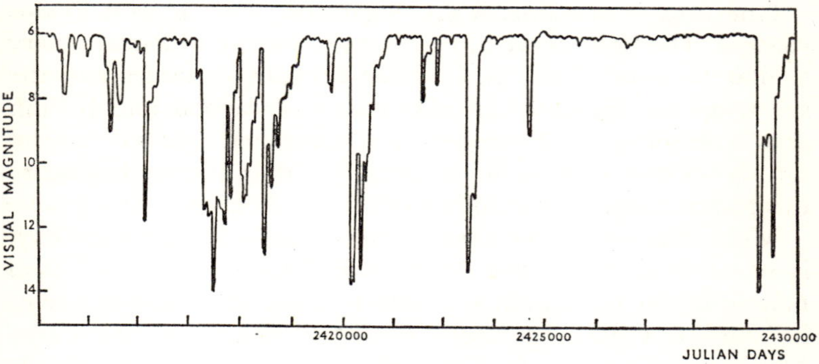

Fig. 51 – Light curve of R Coronae Borealis (*Courtesy of the British Astronomical Association*).

variables. No trace of periodicity exists for these stars and not even a mean cycle can be derived as has been done for so many of the semi-regular variables.

At times, the amplitudes are as large as those found for the long period variables. R Coronae Borealis has been known to fade by more than eight magnitudes at the time of a deep minimum, an amplitude similar to that of σ Ceti for example; but it is difficult, if not impossible, to reconcile a pulsatory mechanism with the light changes observed since for such large amplitudes we would expect such pulsations to be relatively stable and exhibit some form of periodicity.

As in so many cases, we must turn to the spectra of these stars to see if they can provide us with a clue as to the nature of these peculiar light variations. One characteristic feature of the spectra, and one which, as we shall see later provides us with this vital clue, is the high preponderance of the rare type R spectra within this group (Table 20). A few variables are known having earlier type spectra including the prototype star itself which is of spectral class cGOep. The range in effective temperature among these variables is accordingly quite large. RS Telescopii for example has a spectral class as late as R8 which corresponds to a surface temperature of only 2,500°K while at the other end of the scale we have MV Sagittarii, spectral class B, with an effective temperature as high as 15,000°K.

One further variable, V348 Sagittarii, is of interest in this context. The light curve of this star shows several features very similar to those of the R Coronae Borealis variables and the spectrum too indicates an abundance of carbon and helium and a deficiency of hydrogen. The estimated spectral type of O9 shows that this particular star, if it does indeed belong to this group, must be the hottest of the whole class.

Mention must also be made of the variable W Mensae which is situated in the Large Magellanic Cloud. Since its visual magnitude at maximum is about 13.5 we may readily derive its absolute visual magnitude since the distance to the Large Magellanic Cloud is known with a fair degree of precision. This turns out to be −5.0 which agrees well with values between −2.0 and −4.0 for the R Coronae Borealis variables within our own galaxy. From this we may conclude that these stars are all intrinsically extremely bright objects and there are now many reasons which lead us to believe that most of these variables within the galaxy have already been discovered. A total figure of about a hundred of these galactic R Coronae Borealis variables has been estimated.

The main feature of the spectra of these stars is the high abundance of carbon and its compounds in their atmospheres and a marked deficiency of hydrogen and it seems quite likely that during their lifetimes a large proportion of their mass has undergone nuclear reaction while at the same time they remain quite hot, bright stars. They are therefore of great importance in the study of the nucleosynthesis of the heavier elements.

From the spectral type at maximum we know that for the

majority, their surface temperatures are of the order of 7,000°K and all show the 'giant' characteristic of very narrow and sharp absorption lines. It is now considered that there is a definite correlation between the presence of carbon and its compounds

Table 20
R CORONAE BOREALIS VARIABLES

Star	Maximum	Minimum	Spectrum
S Aps	9.6	15.2	R3
V352 Aql	13.1*	(15.0	—
UV Ces	11.8*	16.5	R0?
UW Cen	9.6*	(16.0	K
DY Cen	12.0*	(16.4	—
V504 Cen	12.0*	(18.0	R0?
BG Cep	13.4*	14.2	—
CC Cep	14.5*	16.0	—
AE Cir	12.2*	16.4	—
WX CrA	11.0*	(16.5	R5
R CrB	5.8	14.4	cG0ep
W Men	13.8*	16.0	—
VY Mon	13.7*	15.9	—
Y Mus	10.5*	12.1	R0?
RT Nor	11.3*	16.3	R3?
RZ Nor	11.1*	14.9	R2?
SV Sge	11.6*	13.9	R2
RY Sgr	6.0	14.0	cG0ep
GU Sgr	11.0*	13.8	—
V348 Sgr	11.0*	(16.5	O9
V618 Sgr	11.0*	(16.5	—
SU Tau	9.5	16.0	cG0ep
RS Tel	9.3*	14.6	R8
VZ Tel**	11.6	14.1	—
RZ Vul**	13.0*	(16.0	—

* Photographic magnitude
** Probable members of this class

and the typical light fluctuations since the surface temperatures generally lie within the region where carbon will condense into the solid state. It must not be supposed from this however that such a condensation has the effect of producing clouds of carbon particles of a similar nature to soot although this view is held by some astronomers. One has only to look through the exhaust fumes from a diesel engine to realise how effective ultramicroscopic particles of carbon are in absorbing visible light. Indeed, these extremely small carbon particles are one of the most effective light absorbers known.

The mechanism whereby the small temperature changes take place in these stars thereby producing these highly absorbing layers is not yet known for certain. Two possibilities which have been considered are:

(a) that these clouds of minute particles are formed within the outermost layers of the star's atmosphere and

(b) that the star ejects matter into space at irregular intervals in the form of a circumstellar shell which eventually contracts thereby increasing in density and allowing the deposition of the carbon particles.

It has also been suggested at various times that these stars may be passing through obscuring clouds in space and the accidental occultations produce the observed light changes. This is by no means completely out of the question since many stars are known, including a large number of variables, which are deeply embedded in dark nebulosity; for example the RW Aurigae and T Orionis variables and in their case there is strong evidence that the totally irregular light fluctuations are directly associated in part with this attendent nebulosity. There is one criticism which may be levelled against this argument however. The nature of these dark nebulosities is fairly well known and so far none have been found to contain such a high percentage of carbon.

To examine these possibilities further we must now look at the changes which take place in the spectra of these variables in more detail.

The Spectra of the R Coronae Borealis Stars

As we might expect, the star which has been most completely observed, visually and spectroscopically, is R Coronae Borealis itself. Since the other stars of this group behave in a similar manner we may take this star as being representative of the whole class.

When at maximum brightness, R Coronae Borealis has a

spectrum which is very similar in many respects to that of a supergiant star of type F7 or F8, the major differences being the abnormally weak Balmer lines of hydrogen and the strong lines due to carbon. The identity of a large number of the lines in the spectrum has not yet been established although it appears reasonably certain that they represent transitions within the carbon atom still unknown in the laboratory. The absorption bands due to the CN molecule are quite distinctive.

It is during the fall to minimum however that radical changes occur in the spectrum. These were first noticed more than forty years ago by Joy and Humason during the deep minimum of 1923 when emission lines due to certain ionised metals, notably scandium, titanium and strontium, were observed, these showing a Doppler shift towards the violet corresponding to a velocity of approach of about twenty-two kilometres per second.

About a quarter of a century later, in 1949, when the star faded to a further minimum, additional information was obtained, principally by Herbig, who investigated the changes which took place in the absorption lines. These faded appreciably as the star declined in brightness and were replaced by an emission spectrum which included a bright line due possibly to neutral helium. We may here compare this with the appearance of the emission line of hydrogen which accompanied the deep minimum of ρ Cassiopeiae, a star which in many ways shows a resemblance to the R Coronae Borealis variables and which has indeed been classed as such a star on various occasions. From their spectra it would appear that although both stars have a large circumstellar envelope, that of ρ Cassiopeiae consists of neutral hydrogen while the envelope surrounding R Coronae Borealis and similar variables is essentially neutral helium.

It will be appreciated from the foregoing account that the changes in the spectrum of R Coronae Borealis during the periods of light fluctuation are extremely complex; fortunately it is possible to distinguish three components whose variations appear to be relatively independent of each other and which may be ascribed to different functions of the star's atmosphere or the surrounding gaseous envelope.

(a) The absorption spectrum which is always visible but which undergoes veiling as the star fades, the broad absorptions due to molecular carbon and some hydrocarbons growing gradually more intense towards minimum. The absorption lines themselves show a displacement towards the violet indicating a radial velocity of approach of about twenty-four kilometres per second. Even at a deep minimum the absorption spectrum, particularly in the blue region, is very little different to that shown at maximum brightness.

(b) An emission spectrum composed of sharp, bright lines of neutral and ionised metals together with lines due to carbon and the CN molecule. The latter are of particular importance since, apart from the emission lines of aluminium hydride and oxide which have been seen at certain faint maxima of a small number of long period variables, these are the only molecular lines which have ever been observed in emission. This sharp emission spectrum first appears as the star begins to fade and increases in strength during the initial drop to minimum, growing weaker as the star fades below about tenth magnitude. There is no reddening of this component of the spectrum as the star declines as in the case of the absorption spectrum, nor is any veiling apparent. From this it is concluded that the sharp emission spectrum is produced by the chromosphere of the star as it expands, the rate of expansion decreasing as the star approaches minimum.

(c) The very broad emission spectrum which also has associated absorption lines clearly identified with it. Again, this type of emission appears in the spectrum shortly after the initial drop in brightness but unlike (b) the intensity of these broad lines remains virtually constant throughout the entire period of the minimum, only fading slowly as the star begins to brighten. These lines have been identified as those of most of the metals present in the sharp emission spectrum although it is still doubtful if a broad emission line due to the CN molecule is present and that of molecular carbon would certainly appear to be absent in this component. Since there are no forbidden lines appearing in this broad emission spectrum we may conclude that the density of the material in which these lines originate is comparatively high. From the large radial velocities measured for these lines, about −250 kilometres per second and the fact that these features appear about twenty or thirty days after the initial fading, it is believed that they are due to matter which is ejected at a high velocity through the circumstellar envelope, the atoms being excited as they collide with the gases within the surrounding cloud.

The only direct method of determining the velocity of escape for a star depends upon an accurate estimate of its mass and since all of the R Coronae Borealis variables are single objects, this cannot be determined with the same degree of accuracy as for the binaries. It is generally assumed that R Coronae Borealis has a mass between one and ten times that of the sun. Even if we accept the higher figure it is immediately obvious that the outward streaming gases which produce the broad emission spectrum are travelling far in excess of the escape velocity and do not form part of the stellar chromosphere.

Composition of R Coronae Borealis

Since the recognition of its large hydrogen deficiency and the high proportion of carbon in its atmosphere, several attempts have been made by spectroscopic methods to determine the chemical constitution of the atmospheric material in this star. There are several difficulties associated with such methods and so far no unique determination has proved possible. The lines due to oxygen and nitrogen, for example, are not easy to measure and those due to helium which would seem to form the greater part of these hydrogen-deficient stars are not readily observable in the accessible spectrum.

In spite of these limitations, it has been found possible to make certain measurements of the abundance of some of the chemical constituents. The absolute strengths of the absorption lines due to metals are very similar to those in late F type supergiants despite the fact that hydrogen is essentially absent and from this we may assume that the fraction of material in R Coronae Borealis made up of the metals has not changed to any marked degree during the course of its evolutionary development.

In his spectroscopic investigations, Searle has compared R Coronae Borealis with δ Canis Majoris, a normal F type supergiant situated in a low galactic latitude. This latter star is considered to be typical of the Type I Population, that is a relatively young star. R Coronae Borealis however is believed to be situated at least 3,000 light years above the plane of the galaxy and a statistical study of this class of variable has shown that their spatial distribution and velocities are those of intermediate Population II stars. From the latter we may assume that their masses are approximately the same as that of the sun and if they obey the mass-luminosity law we obtain a value for their mean absolute magnitude of -5.0 which is in good agreement with that obtained by other methods.

To return to the chemical composition of the material making up R Coronae Borealis, Searle has shown that the ratio of carbon to metals in this star is about twenty-five times greater than that found in δ Canis Majoris. Furthermore, from the general weakness of the hydrogen absorption lines in the spectrum, the ratio of carbon to hydrogen may be determined and we find that the outer layers contain at least ten times as many carbon atoms as hydrogen atoms. This marked deficiency is confirmed by the fact that whereas emission lines due to metals appear in the spectrum when the star is below maximum brightness, no emission lines of hydrogen have ever been observed under any conditions.

We are therefore presented with a picture of R Coronae Borealis as a star in which the proportion of metals is virtually the same as that present in a normal F type supergiant but in which a pronounced abundance of carbon is matched by a corresponding deficiency of hydrogen. The bulk of the material is thought to consist of helium, oxygen and neon in that order. So far, we have made no mention of nitrogen. This is certainly present in the variable but estimates have indicated that although it too is probably over-abundant in this star, it is less so than carbon. Owing to the difficulties of observing these latter elements in the spectrum, we have so far no accurate figures for the ratios of each in the atmospheres of this class of variable.

Returning to the question of the condensation of carbon either within the star or at some distance from it, let us see how these two possibilities fit in with the observed facts.

If the condensation occurs within the stellar atmosphere, that is within the photosphere, there should be a differential reddening of the star between maximum and minimum light of about three or four magnitudes between the wavelength limits of 5,900 to 3,550 Å since the ultra-violet flux is very temperature sensitive and unit optical depth in this region corresponds to the higher, and therefore cooler, levels of the photosphere, more so than optical unit depth in the visible region. The result obtained observationally is that although reddening in this region does occur the amount is much less than theoretical amounting to only about one magnitude.

Coming now to the possibility that the carbon condenses in a cloud at some distance from the surface of the star, the question arises as to its mode of formation. Is it formed as a spherical shell expelled from the star, followed by a contraction of this shell after it has been produced, or is it ejected in some preferred direction, rather like a solar prominence only on a far greater scale?

A lot depends upon the dimensions of the chromosphere in these variables. If, as appears likely, they are extensive, then the prominence theory is possibly the more probable. Some support for this idea also comes from the fact that a large number of hydrogen deficient stars containing an abundance of carbon are known but only a small proportion of them show the R Coronae Borealis type of light variation, suggesting that ejection takes place along some preferred axis which may be the result either of rotation or of a magnetic field which has also been shown to be present in these variables.

We must also consider the probability that variability of light in the hydrogen deficient stars may be a comparatively rare (and temporary) phenomenon. The number of minima which have been observed over long periods for these variables differs considerably from one star to another. The longest period during which R Coronae Borealis has remained at maximum brightness since its discovery is nine years whereas MV Sagittarii has dipped to a minimum on only one occasion in the last half century.

From the changes which take place in the spectrum, the presently accepted theory is that during the phase of normal maximum brightness there are turbulent motions within the envelope of the star while at the same time matter rises from the stellar surface with sonic velocities. The onset of a minimum is usually preceded by a small decline before the steeper drop occurs, envisaged as a fall in the level of the photosphere which results in an increased density in the photospheric levels and the deposition of carbon which is the obscuring agent. Some days after the onset of a deep minimum the chromospheric layers begin to rise above the settling photosphere yielding the bright, sharp-lined emission spectrum. As the star fades, the collapsing layers of the photosphere give rise to the expulsion of high velocity gases which move outward from the star and eventually impinge upon the circumstellar material which lies at a comparatively large distance from the surface, the interaction between the gases in this circumstellar cloud (possibly formed during previous minima) and these high velocity particles giving rise to the broad emission spectrum. It is interesting to note that RY Sagittarii also shows evidence of rising material during the later stages of a minimum although this variable is atypical of the class since it shows strong absorption lines of helium but not those of carbon.

We have already said that there are one or two of these stars which have very early type spectra, notably MV Sagittarii (Ble) and V348 Sagittarii (O9e), the latter being also rich in carbon. These stars have presented astrophysicists with a problem which has still not been answered satisfactorily, namely how solid carbon can be formed at the very high temperatures which must prevail in their atmospheres.

Evolution of the R Coronae Borealis Variables

We now come to a very important problem namely the origin of these peculiar stars which as we have said belong to the intermediate Population II stars, and their subsequent evolutionary path. Let it be said at once that it is only possible to speculate on this at the moment. Far more information, both visual and spectroscopic, is needed before we can have any clear idea of their evolutionary status. The R Coronae Borealis stars present so many anomalies that it is difficult even to know where to begin.

The very pronounced deficiency in hydrogen which is the major characteristic of all these stars is also one of the main stumbling blocks to any hypothesis. Were these stars as deficient in hydrogen when they were originally formed out of the pre-stellar gas cloud – or were they initially stars having the normal content of hydrogen which, by some means as yet unknown with any certainty, was used up extremely rapidly during the early stages of their evolution?

Both of these possibilities are quite plausible although from what we know of the composition of the original galactic gas, the latter would appear to be the more probable. The primeval gas from which the galaxies formed was also pure hydrogen and as a result the Population II stars are hydrogen-rich bodies. We must therefore assume that if the R Coronae Borealis variables were formed with this lack of hydrogen they must have done so in regions practically devoid of this gas, regions which had however, for some reason, become enriched with helium and heavier elements.

On the other hand, if R Coronae Borealis began its life as a star much more massive than the sun, then once a large helium core had been formed in the usual way, it may have thrown off the outer hydrogen envelope and what we are now observing is the bare helium core.

Bidelman has suggested that such stars may even be the fore-runners of the planetary nebulae. These are small, compact nebulae, clouds of luminous gas which in a small telescope appear as greenish discs rather like those of the outer planets Uranus and Neptune, hence their name. At the centre of these shells of gas a faint star is almost invariably found and since measurements of their diameters have shown that they are slowly expanding, it is highly probable that the central star, which must have a temperature somewhere in the region of 100,000°K, is responsible not only for their luminosity but also their formation. One of the best known of these nebulae is the famous Ring Nebula in Lyra, easily seen in a small instrument and resembling a smoke ring. It is not, of course, a ring but a slowly expanding sphere. The argument against such a hypothesis is that of all the planetary nebulae which have been examined spectroscopically, every one has a composition very like that of the sun. If any had been formed from the R Coronae Borealis stars we would have expected them to be relatively rich in hydrogen.

Bidelman's suggestion was based on the similarity in the spatial distribution between the planetary nebulae and the known R Coronae Borealis variables. If there is any real correlation between these two classes of object then perhaps we may be more justified in taking the opposite view that the variables have condensed from the nuclei of the nebulae.

Now there is every reason to believe that the lifetime of a planetary nebula is about 50,000 years and therefore it follows that the period in the sequence of evolution in which the star is hydrogen deficient and carbon rich will be correspondingly less, probably only about 1,500 years. As we can see, this is a very short period in the life history of a star and we would anticipate that on this theory any evolutionary changes which are occurring should be fairly readily observable over the course of a century or so.

Have any such changes been observed? Certainly none have been noticed in the case of the large majority of these variables but one star does appear to be showing reasonably pronounced changes in its light curve. This is RY Sagittarii which, apart from the usual irregularities associated with R Coronae Borealis stars, also exhibits a peculiar, pseudo-Cepheid-like variation in a period of about thirty-nine days.

In the earlier chapter on Cepheid variables it was shown that during their evolution these stars pass through the Cepheid instability zone on their way back towards the main sequence (Fig. 19). If now an R Coronae Borealis star were to cross this zone then it too would partake of the familiar Cepheid-like light variations in addition to the more usual irregular behaviour. This is exactly what RY Sagittarii seems to be doing. Certainly the observed amplitude of approximately 0.3 magnitude is far less than that of a typical Cepheid with a period of thirty-nine days and so far the only reason advanced for this anomaly is the loss of the hydrogen convection zone in this particular star.

18. UV Ceti Variables

It will have been noticed from the foregoing chapters of this book that those stars which are variable in light are either binary systems in which the components mutually eclipse each other thereby causing abrupt, temporary decreases in brilliance which are, generally speaking, strictly periodic; or they are intrinsically variable objects where at least the entire outer layers of the star contribute to the variations in light.

However we must not conclude from this that we have exhausted all possible causes of light variability. Since 1947, several stars of low intrinsic luminosity have been discovered which exhibit abrupt flarelike outbursts very reminiscent of the solar flares which have been observed for many decades but of somewhat larger magnitude. The sun is a fairly luminous star and even the brightest flare produces a negligible change in brightness but all of these flare stars are red dwarfs of spectral class dMe showing bright emission lines of hydrogen and ionised calcium in their spectra and these flare-like eruptions are quite sufficient to bring about a fairly dramatic increase in their brightness.

Let it be noted at the onset that visual observation of these flares is neither easy nor unduly rewarding. The variation in brightness of these stars is of such a nature that long and continuous series of observations are necessary to obtain relatively meagre data. It is possibly for this reason that only a small number of astronomers have devoted sufficient time to a systematic study of these stars. Here then is a field of study particularly well suited to the amateur although it must be remembered that, being red dwarf stars, all are very faint and fairly large apertures are required. Later in this chapter we shall discuss the way in which this visual work is being correlated with radio observations of these variables.

At the moment the number of known or suspected flare stars is quite small and when we consider that approximately eighty per cent of the main sequence stars in the galaxy are red dwarfs, this would appear surprising. However the reasons for this are not hard to find. In the first place, being so faint intrinsically, we would expect to observe flares only on those stars which are relatively close to the sun and secondly, the change in brightness during the vast majority of these flares is only of the order of a quarter of a magnitude or so; only rarely are outbursts of two magnitudes or more observed.

The first of these variables to be discovered, quite by accident as it turned out, was the red dwarf known as L-756-8, now designated as UV Ceti. In December 1947 Carpenter made a series of 4-minute exposures of this star on the same photographic plate for the purpose of determining its parallax since it was known to possess a large proper motion (we now know that it is actually the fifth nearest star to the sun). Examination of the plate immediately showed that during the short period the star had been observed (little more than twenty minutes), an unusual increase in brightness of almost two magnitudes had occurred. In many ways the light curve was very similar to that of a nova except that the amplitude was much smaller and the rise to maximum far swifter and it was apparent that an entirely new kind of variable star had been discovered. As always, this discovery prompted further study of the light variations of this star and also a search for similar activity on many other red dwarfs known to lie in the vicinity of the sun. The results obtained over the past two decades, although not as full or as systematic as might have been hoped, justified the initial expectations. More than twenty UV Ceti variables have been discovered in the solar neighbourhood and our knowledge of the physical processes taking place within their atmospheres has been greatly increased by simultaneous visual, spectroscopic and radio studies of these stars. On at least one occasion, in September 1952, UV Ceti increased from its normal magnitude of 12.9 to slightly brighter than sixth magnitude, remaining at this maximum for almost five minutes before fading again. This 250-fold increase in brightness is admittedly unusual for these variables but is an indication of the tremendous activity going on in the atmospheres of these stars.

As with many such accidental discoveries in astronomy, examination of older records revealed that a flare of one of these stars had been noticed as long ago as 1924 when Hertzsprung observed that a faint star in the constellation of Carina had experienced an abrupt rise of two magnitudes on January 29th of that year. After ruling out the possibility that the star was an RR Lyrae variable of ultra-short period, Hertzsprung suggested that the light increase may have been due to a body the size of a planet falling into the star. It is unfortunate that Hertzsprung's discovery was not followed up; instead it appears to have been completely forgotten until the discovery of UV Ceti more than two decades later.

Another flare star of particular interest is Proxima Centauri since this is actually the nearest of all the stars to the sun, lying at a distance of only 4.2 light years. Owing to its close proximity to the first magnitude star α Centauri, it is by no means an easy object to observe particularly with smaller instruments but in spite of this inherent difficulty several brief flares have been recorded, the rise often being in excess of one magnitude.

In addition to the more spectacular flares themselves, photoelectric observations have established other, much slower light variations. Here the increase in brightness is only of the order of a few tenths of a magnitude and the rate of change is about three hundred times slower than that observed for the eruptive outbursts. The present available data is not sufficient for us to say whether there is any correlation between the amplitude of these slow variations and their duration although it can be said that they last far longer than the major flares, and their curves show that the variations are completely irregular. It would appear therefore that a different mechanism is responsible for the two types of light variation noted in these stars.

It is too early yet to carry out a thorough statistical analysis of the two kinds of light variation; many more observations of these stars are needed before this can be done with any hope of obtaining meaningful results. Nevertheless, two conclusions may be drawn from the data we have at present. Firstly, where the increase in brightness exceeds 0.05 magnitudes per second, irrespective of the resulting amplitude, the outburst is of a flare-like character, all others belonging to the secondary changes and secondly, the steepness of the rise during a typical flare is the greatest known for any type of star including the novae.

Table 21
FLARE STARS

Star	Maximum	Minimum	Spectrum
YZ CMi	10.3	11.6	dM4.5e
DH Car	13.1	14.9	dM4.5e
V645 Cen	10.3	11.3	dM5e
DO Cep	9.9	11.4	dM4.5e
UV Cet	5.9	12.9	dM5.5e
EV Lac	8.2	10.2	dM4.5e
AD Leo	9.0	9.5	dM4e
V371 Ori	9.9	11.7	dM3e
EQ Peg	11.0	11.3	dM4e
V1216 Sgr	10.1	10.5	dM4.5e
WX UMa	13.1	14.8	dM5.5e
BD +1° 1522	8.8	9.5	dK2e
BD +33° 1646	10.9	11.1	dMe
BD +51° 2402	5.8	8.3	dK6e
BD +55° 1823	9.6	10.1	dM1.5e
BD −21° 6267	9.8	10.2	dM4.5e
BD −32° 16125	12.1	12.5	dM3e
Ross 154	10.0	10.5	dM4.5e
Ross 867	12.4	13.4	dM5e
Wolf 47	13.1	13.7	dM3e
Wolf 359	12.5	13.5	dM6e
Wolf 424A	10.7	12.7	dM5.5e
Wolf 1130	11.6	12.2	dM3e

All of the above magnitudes are photographic

Table 21 lists most of the flare stars known or suspected at the present time.

At this point it will be helpful to mention briefly the phenomenon of solar flares, since, as will be shown, these are similar in many ways to those thought to be responsible for the major outbursts of the UV Ceti stars. It is now more than a century since Carrington and Hodgson observed a series of very bright patches near the centre of a sunspot group and correlated these with a weak magnetic disturbance on Earth that occurred almost simultaneously and the violent magnetic storm that broke out over fifteen hours later and lasted for almost a week.

This was the first time there had been any direct correlation between a solar flare and magnetic disturbances within the Earth's atmosphere and the measurements of two distinct types of activity were proof that at least two kinds of radiation from the sun were responsible. We now know that whereas the weak disturbance registered within minutes of the outburst on the sun is due to radiation which travels at the velocity of light while the intense magnetic storms are caused by slower-moving radiation in the form of electrons travelling about fifty per cent faster than the speed of sound, around 1,000 miles per hour.

Our knowledge of the true nature of the solar flares is still extremely meagre. Clearly they form very quickly and are of short duration, lasting for only a few minutes before they vanish completely. They are always closely associated with sunspots or with the region on the sun's surface between such spots and appear to be much brighter than the background disc of the sun. Since we know that sunspots possess extremely strong magnetic fields this explains the sharp bursts of radio emission which have been detected from such flares during recent years. In addition to light and radio waves, a typical flare also emits intense ultra-violet radiation of very short wavelength and also X-rays.

Returning now to the flare stars if we assume that these brief rises in light are due to a similar mechanism (and the visual and spectral changes are indeed very similar), then it is evident that to explain the occasional amplitudes of more than three magnitudes such flares must occur on a tremendously greater scale than those in the solar atmosphere. It is true, of course, that all of these variables are intrinsically very faint objects and also relatively close to the sun and although the energy emitted during a single outburst must be several hundred times greater than that of a solar flare such values are not improbable.

The importance of these stars is that investigation of the flares, not only visually but spectroscopically and by means of radio telescopes, gives us vital information regarding the physical conditions prevailing in the atmospheres of these dwarfs especially the radiative mechanism. Since, as we shall see, there is much evidence that these stars are also members of large associations, similar to the T Tauri and T Orionis variables, they are also important from the point of view of cosmogony.

The light curve of a typical flare is shown in Fig. 52.

From continued visual observation it may be possible to

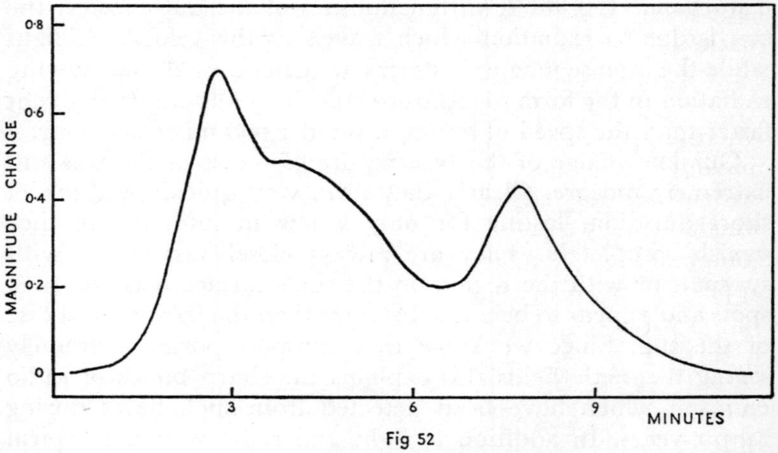

Fig. 52 – Light curve of AD Leonis.

establish some form of cycle for these stars comparable with the sunspot cycle of about eleven years. As yet there is no conclusive proof of this but with further use of the larger telescopes and technical improvements in radioastronomy we may confidently expect even more spectacular advances in this field of variable star research.

An examination of the light curves for many of the flares which have been studied has revealed several interesting features, some of which may be real while others are undoubtedly spurious effects. As has already been stated, very few precise photometric observations of these outbursts have been obtained, the majority being purely visual estimates which, although sufficient to define a light curve, are not always of high enough precision to show any minor irregularities which may be present, particularly on the descent to minimum.

Accepting this limitation, it does appear that all of the light curves are of very similar shape, consisting of the sharp rise to maximum, in some cases preceded by a small increase in brightness observed just prior to the main flare. The decline to minimum always begins immediately after the maximum luminosity has been reached and in general this decline has an exponential form which is quite characteristic. Many attempts have been made to define the fall to minimum by means of a mathematical formula of the kind

$$I = I_0 e^{-at}$$

where I_0 is the peak intensity, t the time in minutes after maximum and a is a coefficient which, from observational evidence, is not constant but may be defined as a *decay function* which, according to Oskanjan who has carried out a very comprehensive and detailed investigation of these variables, appears to be inversely proportional to the maximum intensity I_0.

Spectral Variations of Flare Stars

When discussing the spectra of these variables we must differentiate between those taken when the star is quiescent and those obtained during one of the flare-like outbursts. The former are, of course, comparatively easy to obtain apart from the general faintness of these stars at minimum but the latter present certain difficulties for the astronomer. Fairly long exposures are necessary to obtain satisfactory spectra and unfortunately these flares last only for a matter of an hour or so during which time the light changes are both rapid and varied. We can therefore obtain only an average spectrum of the variations taking place during a flare.

At minimum light all of these variables have emission lines of hydrogen in their spectra and it is generally accepted now that the presence of these lines in the spectrum of a red dwarf star is evidence that it belongs to the UV Ceti class of variable. In addition to these lines, those of ionised calcium are also present.

The intensities of these emission lines vary from star to star and in one of these variables, AD Leonis, which possesses a very extensive Balmer series of hydrogen, emission lines of helium have also been recorded.

A large number of red dwarf stars are known which normally do not show these emission lines in their spectra although they have been noticed on certain occasions and it appears logical to assume that flare-like eruptions are potentially possible with all red dwarfs. We must however distinguish between those in which the mechanism responsible for these flares is permanently present (those which always show emission lines in their spectra) and stars in which this mechanism is active only occasionally (stars whose spectra are normally devoid of emission lines).

When we come to consider the spectral changes accompanying an outburst our information is based on a very small number of spectrograms indeed, all of which however were taken when there was a definite increase in brightness. Unfortunately, owing to a lack of correlation between the changes in the spectrum and precise photometric observations of the light variations, it is extremely difficult to decide whether the brightness changes were of the eruptive type or merely the slower variations already mentioned.

Even from the scanty data which is available however it is possible to differentiate two apparently distinct series of spectroscopic changes which occur in the red dwarf spectrum during the light fluctuations. In the first an intense ultra-violet continuum is superimposed upon the spectrum almost completely blanketing the absorption features in this region. At the same time the hydrogen emission lines increase in intensity although those due to ionised calcium remain virtually unaltered. Emission lines of neutral and ionised helium have been observed on one or two occasions.

In the second series of changes which have been observed, there appears to be no variation whatsoever in the continuum. The Balmer lines of hydrogen normally present in the spectrum increase somewhat in strength and lines of a higher order make their appearance. Again, there is very little change in the H and K lines of ionised calcium.

It would thus appear that here we have spectroscopic

evidence for the existence of two kinds of reaction going on within the atmospheres of these red dwarf stars producing the rapid and slow outbursts of light. We must now look more closely at other lines of investigation which have provided further information concerning the nature of these mechanisms.

Polarisation of Light from Flare Stars

The fact that the light from a typical solar flare is partially polarised has been known for some time and if similar flares are responsible for the light variations of the UV Ceti stars we would expect evidence of polarisation in their case. Oskanjan has made a series of observations on a fair sample of these red dwarfs and shown that the light from certain of these stars is polarised to a variable extent, the degree of polarisation being greater during the increase to maximum light than at maximum itself. Owing to the paucity of observations it is not yet possible to state that there is a definite correlation between the amount of polarisation and the strength of the hydrogen emission lines although this appears likely. The fact that the light is polarised is important in our understanding of the nature of these flares since it provides proof that they are not thermal in character.

The Nature of the Flares on UV Ceti Stars

Since the discovery of the flare stars two decades ago, many attempts to provide an explanation of the abupt flare-like increases in brightness have been made. Perhaps it is only natural that comparisons were made between these variables and other classes of variable star whose light curves show the same characteristic, namely outbursts of light which are of an eruptive nature; i.e. the U Geminorum, RW Aurigae and T Orionis variables and the normal and recurrent novae. From all that has been said however it is reasonably clear that the outbursts occurring in the flare stars are in a class on their own. Not only are the spectral changes unlike those of the other variables but the duration of a typical flare and the rapidity of the rise to maximum are of a different order altogether.

Analogies have also been drawn with the solar flares and here the similarity is a little more definite. A solar flare is of about the same duration as a UV Ceti type outburst and the light polarisation is of the same order. The general shape of the two light curves are also very similar and in both cases emission lines of hydrogen are strengthened in the spectrum during an outburst. The major difference between the two phenomena is that whereas emission lines are visible in the spectrum of a flare star at all times, none are present in the solar spectrum except during a chromospheric flare.

It was initially considered that the flares in the UV Ceti variables had a thermal origin but more recently much evidence has accumulated against this idea. Neither the polarisation of the light, nor the incredible steepness of the ascending branch of the light curve can be reconciled with any type of thermal radiation. We must also take into account the fact that most of these outbursts last for only a matter of minutes rather than hours and not only would the transmission of thermal energy have to be effected within seconds but it is inconceivable that a large mass of gas could cool down by more than two thousand degrees in a matter of ten or fifteen minutes as is actually observed. The final evidence against the theory of a thermal mechanism is provided by the spectrum. Here the energy distribution in the continuum is quite different from that we would expect from a mass of gas suddenly heated to some thousands of degrees above its normal temperature. We must therefore look for some other atmospheric disturbance in these red dwarfs which is capable of producing such phenomena.

It must be emphasised yet again that far more observations will be needed before we have a reasonably clear picture of this phenomenon but the suggested mechanism put forward by Oskanjan appears plausible in the light of our present knowledge. On his theory, a large release of energy from the deeper regions of the star is followed by its emergence through the upper levels where it is transformed by a non-thermal process into radiation. The slow variations in light are produced whenever the process occurs in the deeper levels of the star's atmosphere, the more violent outbursts taking place when the release of energy takes place within the higher levels.

As we have seen, in many cases there are marked irregularities on the light curve and it is probable that in some instances at least, these outbursts take the form of jet-like prominences rather than a continual steady flow of energy from the star although this is postulated in those stars in which bright emission lines are always present.

Spatial Distribution of Flare Stars

The point has already been made that the percentage of red dwarfs among the main sequence stars is very high and it is therefore only to be expected that fairly large numbers should exist within fifty light years or so of the sun. Since they are intrinsically faint stars, almost all of the UV Ceti variables so far discovered are close to the sun, making it a relatively easy matter to determine their proper motions and hence their actual spatial velocities. One important result which has emerged from this work is that those red dwarfs having emission lines in their spectra (and which are therefore either known or potential UV Ceti variables), have a smaller dispersion of spatial velocities than those which show no emission lines suggesting strongly that the flare stars may be members of an association similar to the T Tauri stars which have already been described.

Further evidence in favour of this comes when we examine the distances of the UV Ceti variables. Within a sphere of radius one hundred light years no UV Ceti stars have been found beyond sixty-five light years, the numbers decreasing with increasing distance. The question immediately arises as to whether this effect is real indicating the presence of an association of these variables around the sun or whether the result is due entirely to the intrinsic faintness of these variables. Clearly the observed facts will fit either possibility but when taken in conjunction with other evidence the probability is that the spatial distribution is real.

One characteristic of stars which form a stellar association is the high percentage of binaries which exists among them and it is perhaps significant that more than sixty per cent of the known UV Ceti variables are members of binary systems. This raises the point that their duplicity might, in some way, be responsible for the light changes although at the present time there is no evidence to support this. From the foregoing discussion we see that there is a great deal of observational evidence that the sun is close to the centre of a cloud of UV Ceti variables which form a stellar association. The chance of such a clustering of these stars having been brought about by random motions is out of the question and we are therefore forced to the conclusion that most, if not all, of the red dwarfs in the neighbourhood of the sun had a common origin.

There are, naturally, other associations of UV Ceti stars. The study of the T Tauri variables in the constellations of Taurus and Orion has indicated that among these stars there are large numbers of variables which, from their light and spectroscopic characteristics, are clearly flare stars. Measurements of the diameters of these clusters have shown that they are of the same order as that near the sun, namely between sixty and seventy light years.

Radio Emission from UV Ceti Variables

We are now in a position to discuss the evidence for radio emission from these stars. As soon as it was suspected that the observed flares might be of a similar nature to those seen on the sun – and still more when it was realised that, whatever their true nature, the energy emitted must be several hundred times more than that of even the brightest solar flare – the question arose as to whether it might not be possible to detect such radio emission using sufficiently sensitive instruments. If such radio studies could prove the existence of an emission from flare stars then it would add further information about the extent to which the ejection of particles possessing high energy into interstellar space might contribute to both cosmic radiation as a whole and also the radio waves which are emitted by the Milky Way.

For many years, radio astronomers have been engaged in joint programmes with visual observers hoping to correlate visual flares with radio emission. Much of this work has been carried out on southern UV Ceti stars by the Radiophysics Laboratory at Sydney, Australia, particularly on Proxima Centauri, UV Ceti and V1216 Sagittarii, using wavelengths of 3.5 and 15 metres.

At present, the data obtained is not sufficient to give conclusive proof that radio emission has been detected from these flare stars especially since no radio emission was found at the time of two bright visual flares of UV Ceti. However, at the times of three probable flares where the rise in brightness was quite small, of the order of 0.3 magnitude, radio deflections were recorded at 3.5 metres, and on one occasion an extremely strong burst at fifteen metres followed closely on a weaker burst at the shorter wavelength. The observed delay between the two observations was very similar to that often recorded during type II radio bursts observed with solar flares. In every case, the radio emission was shown quite conclusively to originate in the neighbourhood of the flare star.

19. The Novae

With the exception of the supernovae which will be described in Chapter 22, the novae are among the most spectacular of all variable stars. Without any previous warning whatsoever they suddenly flare up from obscurity, increasing extremely rapidly in brilliance until they may even outshine the brightest stars in the heavens. Once maximum has been attained a far slower decline sets in often accompanied by minor fluctuations until, after a period which may last from two to twenty years, they regain their former faintness although frequently continuing to vary in an irregular manner over a fairly small light range.

The term 'nova' meaning quite literally a new star was applied to these objects before the use of photography and large telescopes, in conjunction with accurate star charts, gave a truer picture as to their real nature, namely that they are not *new* stars which have sprung spontaneously into being, but extremely faint objects which experience an incredible explosive outburst that shoots them from anonymity to some of the brightest stars known. The point must be made here however that by no means all of the novae become naked-eye objects at maximum brightness nor, if they are, are they discovered at the time of reaching maximum. Very often a nova may only be discovered many years after the event by a careful comparison of sky patrol plates which are taken at regular intervals and cover the whole of the sky.

It is for this very reason that nova patrols, i.e. sweeping the sky, particularly along the boundaries of the Milky Way where most novae are found, are of great value. Many novae attain at least eighth magnitude at maximum and provided one has a good working knowledge of the stars down to this magnitude, it is not impossible to discover a nova as may be shown by the fact that a large proportion of the brighter ones of the present century were discovered by amateurs, several of whom were actually searching for comets at the time!

One of the brightest novae of recent times appeared in the constellation of Aquila on June 7th, 1918. Patrol plates of the region taken prior to the outburst revealed it as an inconspicuous star of about the eleventh magnitude, different in no way at all from the hundreds of others on the plates. That night it had reached 5.8 magnitude and soared to first magnitude by

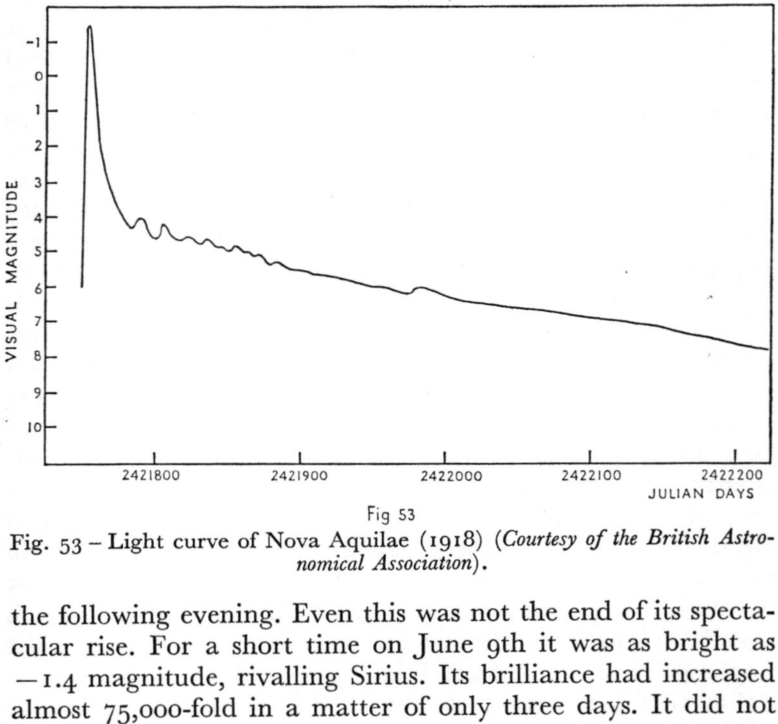

Fig. 53 – Light curve of Nova Aquilae (1918) (*Courtesy of the British Astronomical Association*).

the following evening. Even this was not the end of its spectacular rise. For a short time on June 9th it was as bright as −1.4 magnitude, rivalling Sirius. Its brilliance had increased almost 75,000-fold in a matter of only three days. It did not remain long at this brightness however and by the end of the month it had faded appreciably to somewhere between third and fourth magnitude and at this point the rate of decline slowed markedly, the star not reaching its original magnitude until towards the end of 1924. The light curve of Nova Aquilae (1918) is shown in Fig. 53.

The idea of some catastrophic outburst is confirmed by a

spectroscopic examination of a nova. In the case of Nova Aquilae (1918), the spectrum shortly after maximum light revealed broad absorption features which were displaced towards the violet indicating high radial velocities of approach of the order of 800 kilometres per second. Visual proof of this came not long afterwards when the shell of gases expelled from the star were visible in large instruments in the shape of a tiny nebulosity, the diameter increasing perceptibly as the star faded. As time went on this nebulous shell became so faint that it was almost impossible to observe visually but as late as 1941 it was successfully photographed by Baade in red light who used the same technique to show the presence of similar gaseous envelopes around some of the earlier novae, in particular Nova Aurigae (1892) and Nova Cygni (1920).

Now if we proceed on the reasonable assumption that the radial expansion of the gases expelled from Nova Aquilae (1918) as given by its spectrum, was initially about 1,000 kilometres per second, this will give us an approximate distance to the nova of around 1,000 light years and an absolute magnitude of −9.0 at maximum and about +2.9 at minimum. We must however treat these values with a certain amount of caution since later work has shown that in a large number of cases the gases which are thrown out by a nova explosion do not move in a uniform manner through space but appear to be ejected in two diametrically opposite directions. Fortunately in the case of this particular nova it has been found that there is a coincidence between these favoured directions and the line of sight as seen from Earth but in other instances, notably Nova Herculis (1934), discovered by J. P. M. Prentice, by far the greater proportion of the gases were expelled sideways, at right angles to our line of sight and the above assumption is no longer valid. In the telescope, this has had the effect of making it appear that there are two luminous spots situated on either side of the central star.

Novae are considered to be dwarf stars and at minimum it is estimated that their luminosity is generally comparable with that of the sun; but at maximum it is between 30,000 and 75,000 times brighter. This incredible increase in luminosity is far in excess of anything we have previously considered. If the sun were to turn nova, the Earth and the inner planets would certainly be incinerated and possibly several of the outer planets too! Fortunately for the continuance of life on Earth it is highly unlikely that such a catastrophic event will overtake the sun at least for many millions of years.

The earliest novae, and several have been recorded over a great many centuries, were regarded like the naked-eye comets with awe and superstition. Even during the last century when a little was known of their true nature, the first scientific explanations which were put forward to explain them are known to be completely untenable. For example it was once supposed that the passage of a very faint or dark star through a dense cloud of interstellar matter resulted in an extremely rapid heating of the outer stellar layers to incandescence, rather like a meteor which burns up in its flight through the Earth's atmosphere. This idea received some support from the observation that the great majority of novae appeared along the boundaries of the Milky Way where one might expect to find such cosmic dust clouds. Certainly this provided an explanation of the abruptness of their appearance and the much slower decline in brightness.

Another suggestion was that the nova outburst was the result of a head-on collision of two faint or completely dark bodies, ending in the total annihilation of both, the two stars being vaporised in the process. With the eventual realisation of the tremendous distances involved and the utter emptiness of interstellar space within the galaxy it was seen that the chance of such an encounter was so fantastically remote that it could not possibly explain the number of novae which have actually been observed. Then, too, several novae were found in regions of the sky where there is no evidence for the existence of dark obscuring matter, thereby disposing of the first theory.

A third idea that a large swarm of massive meteorites may have plunged into the atmosphere of a dim star was also discarded when calculations showed that the total mass of such a meteor swarm would have to be impossibly high to produce the observed increase in radiation and temperature. Finally the spectroscope provided the explanation which is now universally accepted: that of an explosive process in the outer layers of the star.

Light Variations of the Novae

It must not be thought that all novae follow the same pattern in their light variations, nor in their spectral characteristics although the latter are certainly more consistent than the former. Strictly speaking, perhaps, each nova should be treated as an individual but from the form taken by their light curves they may be conveniently divided into three general classes depending mainly on the initial rise to maximum and the behaviour of the star during the hundred or so days following maximum brightness. It must be stressed that the divisions between these three classes are by no means sharp and in particular many novae show features which are common to both of the first two groups. In addition, certain of these stars possess features in their light curves which place them outside of these classes. One in particular is Nova Delphini (1967), discovered on July 9th, 1967, by G. E. D. Alcock and which, although still around maximum at the time of writing, promises to be one of the most exceptional of these variables. Another is η Carinae, a star whose light variations are so unique that it will be dealt with in detail in a further chapter.

The three divisions of the novae are given below:

Na: These are the rapid novae which have an extremely steep rise to maximum of a few days, or even a single day, succeeded by an initially sharp decline which later slows and may then become reasonably smooth (Nova Cygni 1920, Nova Puppis 1942) or the fading may be marked by a prolonged series of quite pronounced fluctuations (Nova Persei 1901, Nova V368 Aquilae 1936). The quite arbitrary criterion which has been used to classify novae of this class is that the rapid rise should be followed by a decline of three magnitudes in 100 days or less. The light curve of Nova Cygni (1920) is shown in Fig. 54 and may be compared with that given in Fig. 53.

Nb: The second broad group known as the slow novae are characterised by a more gradual development, the rise to maximum often taking a fortnight or more and the star may remain at, or around, maximum for several weeks or months before the inevitable decline sets in. These novae tend to fade quite slowly at first (the decrease of three magnitudes takes

150 days or more) and there are often very marked fluctuations during this slow decline. A point is reached however when the rate of fading quickens and they fade to something approaching their normal magnitude. They do not, however, always remain at this magnitude but brighten slowly and irregularly to a second maximum from which they do eventually return to minimum. Nova Herculis (1934) is a good example of this type

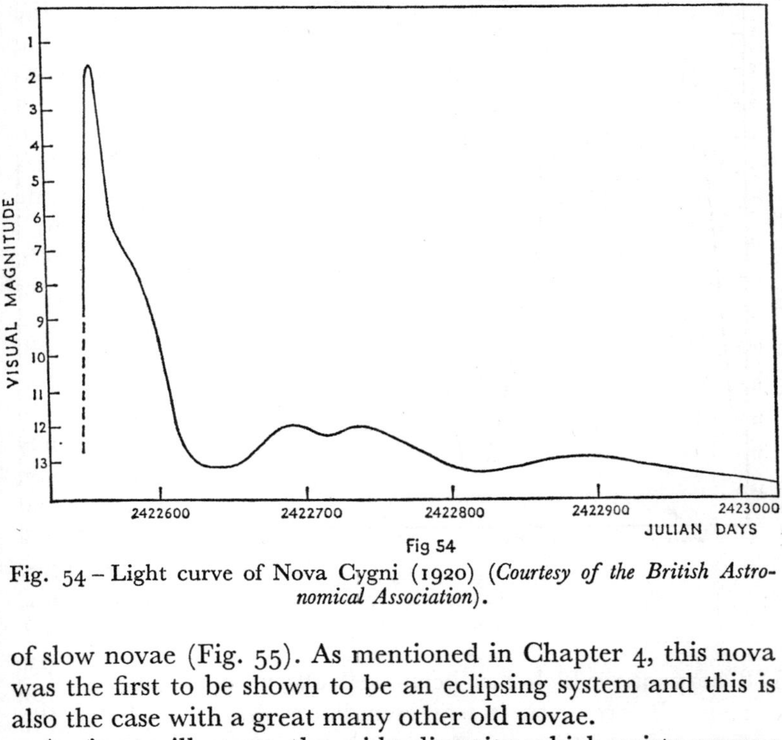

Fig. 54 – Light curve of Nova Cygni (1920) (*Courtesy of the British Astronomical Association*).

of slow novae (Fig. 55). As mentioned in Chapter 4, this nova was the first to be shown to be an eclipsing system and this is also the case with a great many other old novae.

Again, to illustrate the wide diversity which exists among these stars, there are other novae classed in Group Nb, for example Nova Pictoris (1925) and Nova Aquilae (1936 I) which do not exhibit this secondary brightening.

Nc: This is a very small group of ultraslow novae containing only a few members. Here the light curves show many similarities to those of the two preceding groups but the maxima extend not over months, but during the course of several years. The decline also proceeds with extreme slowness. The first such star to come to the notice of astronomers, and by far the most spectacular and puzzling because of its extended activity and the wide range of its fluctuations, is η Carinae. An extensive literature has grown up devoted to this particular variable and it will be treated elsewhere in this book. A further example is

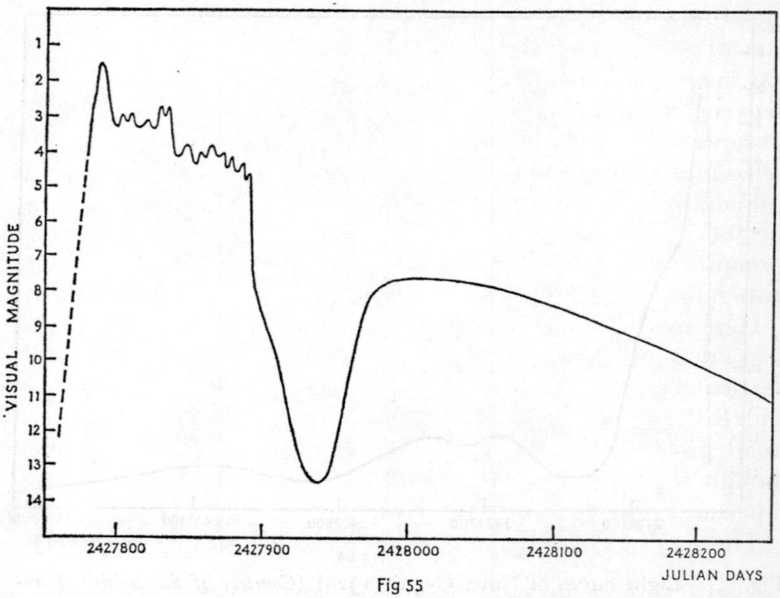

Fig. 55 – Light curve of Nova Herculis (1934) (*Courtesy of the British Astronomical Association*).

RT Serpentis which has been well observed for over half a century. Before 1909 it was completely unknown. From the photographs of the region taken before the outburst, it was certainly fainter than the sixteenth magnitude and beyond the reach of most instruments available at that time. During that year however it rose slowly and erratically to 11.7 magnitude and became readily observable. The slow brightening continued until it attained 10.5 magnitude in 1919, remaining there until 1923. The star then declined gradually in brightness to about magnitude 13.2 where it has stayed ever since. There is no guarantee however that it will not brighten up once more since these stars are quite unpredictable in their behaviour.

A third star which must be considered here appeared in the constellation of Orion in 1939 and is designated as FU Orionis. At maximum it reached 9.5 magnitude being then more than five magnitudes brighter than at minimum. The light variations followed a similar course to those of RT Serpentis and the spectrum of this particular variable has been given as cF5 like that of η Carinae which would indicate that it too is a giant star having a surface temperature at maximum somewhere between 7,000 and 8,000°K. This fact is quite significant since it shows that in this small group of stars where the nova-like behaviour is spread over a large number of years, the surface temperature is substantially lower than that found at maximum for the true novae of classes Na and Nb. Even so, the total amount of energy radiated by these stars must be truly stupendous.

Here it will be worthwhile to anticipate a little of what will be said in the chapter on the supernovae and consider one particular star which bears so close a resemblance to η Carinae and RT Serpentis that it should perhaps be included in this small group of variables we have just been discussing, even though much of the available evidence points to it being a supernova, rather than one of the ultraslow novae.

The star in question was discovered by Wild in 1961 in the outer environs of the spiral galaxy NGC 1058 and it immediately presented certain problems. The light curve bore such a close resemblance to that of η Carinae that at first it seemed very doubtful if it could belong to NGC 1058 and it was thought to be a normal nova within our own galaxy and merely seen projected against the background of the nebula. Examination of earlier photographic plates of the region showed for example that since 1937 the star had exhibited small variations in brightness before rising to magnitude 12.2 in December 1961, after which it faded slowly to 19.0 magnitude by late 1963.

All of the absorption lines in the spectrum of this peculiar star were shifted towards the red with a mean value of about

750 kilometres per second whereas the velocity of recession of the spiral galaxy itself is much less, being only around 450 kilometres per second. So far, it is not possible to give a definite explanation of this large discrepancy in these velocities although there are two possibilities; either the gas thrown out by the outburst was ejected in two preferred directions like that observed in Nova Herculis (1934) and these directions happen to lie along our line of sight or the nova itself is receding from us with a velocity of 300 kilometres per second. The actual velocity of expansion of the nebular shell has been estimated by Zwicky as 3,000 kilometres per second and since no forbidden lines have been observed in the spectrum is appears that the mass of gas ejected is much greater than that expelled by ordinary novae. The absolute photographic magnitude of -15 means that the star was about two magnitudes brighter than η Carinae at its 1843 maximum but fainter than the normal supernovae.

We may now make some interesting calculations about the frequency of occurrence of the nova phenomenon. In spite of the impressive nature of the explosion, very little of the actual mass of the star is concerned in the process, almost certainly not more than ten per cent, almost all of which is hydrogen. Of this ten per cent, only about 0.01 per cent is actually expelled and lost by the star, the remainder being held by the star's gravitational field and as time goes on, this gas will cool and return to something like its former state where it becomes available for a further outburst at some time in the future. When we come to discuss the physics of the nova phenomenon we shall see that this is the means by which the star rids itself of its surplus hydrogen and quite clearly, from what has just been said, it will require something like a thousand outbursts to dissipate the whole of the surplus hydrogen. Not every star, of course, will follow a path of evolution which will result in it turning nova at some time during its career, but it is a relatively simple matter to calculate roughly how many such stars there are within the galaxy and the answer turns out to be about 100,000,000. Accepting this figure, we can see therefore that we arrive at a total of about 100,000 million individual outbursts in the galaxy. To obtain the number of novae which should appear each year we must know the time of evolution of these particular stars which is known, with a fair degree of precision, to be of the order of 5,000 million years. Calculation therefore shows that we would expect to observe about twenty novae each year within our own galaxy.

How does this figure compare with observation? The answer is, very well indeed. Certainly many of the novae are extremely faint objects even at maximum and others are only observed some years after the event by a comparison of sky patrol plates since otherwise they are lost among the thousands of faint stars which border the Milky Way.

There is one further result which we might expect from the fact that a star can experience a nova explosion more than once in its lifetime. The intervals between successive outbursts will, for most of these stars, be measured in thousands of years. But is it possible that there may be some stars, admittedly only a small number, in which this time interval is much shorter, say of the order of twenty to fifty years? Again, observation has shown that this is indeed the case. These recurrent novae as they are termed have been discovered in increasing numbers and are being extensively studied since they provide us with additional information on the nova phenomenon. These very interesting variables will be discussed in Chapter 21.

Detailed investigation has brought to light an apparent relationship between the novae, the recurrent novae and the U Geminorum variables which have been described in Chapter 14. The amplitudes of the U Geminorum variables are less than those of the recurrent novae which in turn are generally smaller than those of the novae themselves. There is also a similar decrease in the period between outbursts and from this it has been suggested that there may be a correlation between amplitude and period giving, by extrapolation, periods ranging from several hundred years to several millennia for the novae. So far, this is only supposition but much of the evidence, although admittedly circumstantial, would seem to be in favour of it.

There is no doubt that with any kind of variable star, interpolation is imprudent and extrapolation an extremely dangerous procedure often leading to completely erroneous results. It is certainly very tempting to accept this theory, particularly since the values obtained appear to fit the facts as we know

Table 22

THE BRIGHTEST NOVAE

Nova	Maximum	Minimum	Spectrum
N Aql (1899)	5.5*	(15.6	Q
N Aql (1918)	−1.4	10.8	Q
N Ara (1910)	6.0*	17.5	Q
N Aur (1891)	4.1*	15.8	Q
η Car (1843)	−0.8	7.9	Q
N Cyg (1876)	3.0	15.2	Q
N Cyg (1920)	2.0	15.4	Q
N Del (1967)	3.6	11.9?	Q
N Gem (1903)	4.8	16.5	Q
N Gem (1912)	3.5	14.7	Q
N Her (1934)	1.3	15.4	Q
N Her (1960)	3.0	variable	Q
N Her (1963)	2.9	—	Q
N Lac (1910)	4.3*	14.4	Q
N Lac (1936)	2.1	15.6	Q
N Lyr (1919)	6.5*	15.4	Q
N Mon (1918)	5.2*	15.1	Q
N Mon (1939)	4.5*	16.0	Q
N Nor (1893)	7.0*	(16.3	Q
N Oph (1848)	4.3	13.0	Q?
N Oph (1917)	6.2*	(17.0	Q
N Oph (1919)	7.2*	(15.2	Q
N Per (1887)	4.0*	(17.5	Q
N Per (1901)	0.2	14.0	Q
N Pic (1925)	1.2*	13.6	Q
N Sge (1783)	6.0	15.8	—
N Sgr (1898)	4.5*	16.5	Q
N Sgr (1936)	4.3	15.2	Q
N Sgr (1941)	6.8*	(16.0	Q
N Vul (1670)	2.7	(17.0	—

* Photographic magnitude

them at the moment. If we are on safe ground in extrapolating in this way, it is to be expected that the novae would be observed only once during the time in which the science of astronomy has been seriously studied and this, combined with the information which has been derived from the recurrent novae, provides us with additional confirmation of the conclusions we have already reached that (a) only a certain type of star experiences a nova outburst and (b) that only the outermost layers of the star are affected during the explosion and expelled as a shell of gas, the star returning to something akin to its former self and able to undergo the entire process many more times.

Table 22 lists the brightest novae which have appeared in the last century or so. Nova Vulpeculae (1670) has been included here because of its brightness at maximum.

Spectral Changes in Novae

Although all novae are far from sharing the same characteristics or exhibiting a like behaviour, certain well defined changes occur in their spectra following the outburst. Since the rise to maximum is exceptionally steep in most cases, it will be appreciated that only on rare occasions has it proved possible to obtain spectra during the increase in brightness.

Before the outburst, the spectrum is normally of class B or A, sometimes as early as late-type O as is probably the case with Nova Delphini (1967). When at maximum brightness, the spectrum is quite characteristic consisting of a bright continuum with the maximum intensity far in the ultra-violet, sometimes with very broad bright and dark bands due to hydrogen from the violet shift of which the velocity of approach of the expelled gases may be measured. Gradually, the lines widen and the continuum begins to fade. All of this time, the star is white in colour but soon it becomes more yellowish and the nebula lines appear in the spectrum. As the nova declines in brightness, it usually becomes more red in colour and the hydrogen lines in emission become more intense, fading later as the colour changes to orange. After a time the nebula lines become more prominent than those due to hydrogen, the colour of the nova being a leaden-blue. Towards the end of the decline the spectrum consists of a faint continuum with bright Wolf-Rayet bands due to atoms of an extremely high excitation potential. The spectrum then is indicative of a surface temperature in the region of 30,000 to 35,000°K.

Naturally, the above is only a general description of the sequence in the spectra and colour of a nova after maximum. Very often there are considerable variations from the scheme just outlined with some stages missing and individual peculiarities peculiar to each nova.

The Physics of the Nova Phenomenon

It is quite clear from what has been said that the origin of the nova outburst must be looked for in some physical mechanism within the star itself and even today we do not know the precise cause although a reasonably clear picture of events taking place in the interiors of these stars can be postulated and we are beginning to have a greater understanding of the nuclear processes which precede the actual explosion. We must now therefore consider the possible causes of instability which can conceivably lead to such a dramatic and catastrophic disruption of the outer layers of such a star.

It is in the centre of the star however where we must look first for the nuclear reactions which may eventually lead to a nova explosion. It is here where temperature and density are greatest, that hydrogen is first transformed into helium resulting in a core of this element. Now unlike hydrogen, the nucleus of helium is extremely stable and hydrogen burning will continue for some time, increasing the size of the core before gravitational contraction occurs, raising the temperature at the centre sufficiently for the stable helium to begin burning, producing the heavier elements such as oxygen, nitrogen and neon.

As will be explained in detail when we consider the supernovae, little if any mixing of the hydrogen and helium takes place within the star with the result that a relatively thin energy-producing skin is formed around the growing core which gradually works its way towards the surface as more and more hydrogen is consumed. Now calculation shows quite conclusively that since the temperature of this skin is somewhere in the region of 10,000,000°K it will be physically impossible for the surface of the star to be as hot as this since the only way by which energy can escape into space from the star's surface is by radiation and the surface area of the star is such that the amount of radiation at this high temperature would be impossibly great. The star would be literally unable to lose its energy at a sufficient rate. It is not merely a case of the radiation being ten or twenty times too great, but several thousands times too high.

Unfortunately for theory, it has been shown that some stars do burn up their hydrogen almost to the surface and we are therefore forced to postulate some means by which this may be done.

To begin with, let us suppose that our earlier statement that no mixing of hydrogen and helium occurs is not strictly true and that owing to the presence of very powerful convection currents in the upper atmosphere it is possible for some of the surplus hydrogen to circulate downward towards the interior of the star. This mixing, limited as it is, would certainly allow some of the helium to come to the surface rather than going to make up the growing core. The chances are that such a process does occur but not to any great extent and almost certainly not to the extent that would be necessary to solve the problem. What is much more likely is that the outermost layers become degenerate in the same way as the helium core as originally suggested by Mestel.

From theory, we can show that in this event an explosive condition arises and a vast flood of energy is released in one of two ways; either as heat which, in the visible region of the spectrum at least would account for the abrupt rise in luminosity as shown by the phenomenal amplitudes of the novae, or if translated into motion would produce an expanding shell of gas, moving outward from the star with incredibly high velocities in excess of 1,000 kilometres per second. Since both of these conditions are observed in the case of novae, it seems likely that the energy is released by these two methods virtually simultaneously.

So we now see that, as opposed to the supernovae to be discussed later, only the small proportion of stellar material in the outer layers is expelled by the star and since both visual and photographic observation has shown that virtually all of the expanding gases are not dissipated into space, then quite clearly their velocity away from the star does not overcome the tremendous gravitational attraction of the parent body. After an interval of some years, as we have already said, the intensely hot hydrogen will cool sufficiently to return to something like its original state.

Close Binaries Among Old Novae

This question has been discussed earlier in Chapter 4 but since the consequences of such binary systems are pertinent to alternative theories of the nova phenomenon a reiteration and extension of the discussion will be given here. So far we have gone on the assumption that a nova is a single star. Since 1954 when Walker discovered that Nova Herculis (1934) is a close binary system, several other old novae have been shown to possess a binary nature. One of the more remarkable features of these systems is that they all appear to consist of a blue star and a red component which is losing mass through the inner Lagrangian point, material which forms a ring around the blue companion and moves with it in its orbit around the common centre of gravity. Although at the moment the amount of available data is admittedly limited, it would appear that this rather strange combination of a blue and red star among the old novae cannot be explained merely by coincidence. As Kraft has suggested, it seems far more likely that this combination is a prime requisite for a nova outburst. Let us now look into the position a little more closely and see whether this situation can provide us with an alternative explanation of the nova phenomenon.

We shall again assume that the star which is the seat of the nova explosion contains only a small amount of remaining hydrogen, this being in the form of an outer atmospheric layer, the remainder of the stellar material being helium which is in a degenerate state. Furthermore, Kraft assumes that it is the blue companion which is the actual nova, an assumption which appears logical although there is no direct observational evidence for this. Then we may envisage a situation where the gases from the red component as well as surrounding the blue star also penetrate the thin hydrogen atmospheric layer into the hot helium core where the temperature is sufficiently high to start the hydrogen burning. Since the helium is degenerate, the result will be an increase in the reaction rate with the star unable to cool by expansion as would be the case if the material were non-degenerate. If then, by some mechanism we do not yet fully understand, the degeneracy could be removed, the star would then turn nova.

Unfortunately, as is often the case, there are serious obstacles to this very plausible theory. Certainly it does fit the observed fact that the necessary stream of hydrogen-rich material is present in the binary system but is it plausible to suppose that the outer regions of this large, degenerate core can be heated quickly to a high enough temperature for them to take part in a nova explosion? All of the indications of the theory of degenerate matter are that this will not be possible since such material happens to be a very good conductor of both electrons and heat. Can we then presume that given a sufficient length of time it will be possible to raise the temperature of the entire helium core of the star to the point where an explosion will occur? The answer appears to be yes. But now we come up against one of the major obstacles to this particular theory. If this happens, then, as we shall see later, the amount of energy released will be so great that a supernova and not a nova explosion will occur. There can be no doubt about this as was shown by Mestel several years ago. At the present time therefore it appears that this theory, in spite of its many attractive points, must be discarded.

There is yet a third theory which we may consider based upon the binary nature of these variables. When we discussed the Algol variables we saw that in certain cases the period of axial rotation of a component could be non-synchronised with the period of revolution of the system and that where this happened, there is almost invariably a gas stream present in the system flowing from one star to the other. Since such gas streams are present among the binary systems of the old novae it strongly suggests that the axial rotation of the blue star is not synchronised with the period of revolution of the red companion. Almost thirty years ago this problem was considered by Cowling who showed that in certain cases non-radial oscillations could be induced into such a star and provided that certain conditions are fulfilled, these may resonate with the orbital period. Applying this theory to the old novae, Schatzman has demonstrated that these resonating non-radial oscillations will eventually lead to material being ejected from the blue star, not as spherically expanding shells of gas but as cones. Now we have already seen that the gases thrown out by a nova appear to be ejected in two preferred directions and here we have a theoretical explanation of this, a very strong point in favour of this theory since neither of the others provides for this observational fact. On one important point, however, the theory does not agree with observation. The nature of the secondary component does not appear to be critical whereas we have seen that all of the available evidence suggests that it must be a red star which has filled its lobe of the equipotential surface and is overflowing through the inner Lagrangian point. We must therefore end this section on the mechanism of the nova phenomenon on a note of uncertainty. All three theories have points in their favour, yet none of them satisfactorily explains all of the observational data.

Galactic Distribution of the Novae

It is a well-known fact that most of the novae have been discovered along the boundaries of the Milky Way and consequently they belong to the galactic disc. They are therefore Population I stars. However, the situation is not quite as simple as this since a small number of novae have also been discovered by Baade in two elliptical galaxies and one nova, T Scorpii (1860), in the globular cluster Messier 80. These latter would therefore appear to belong to the Population II stars.

Now this observation of novae occurring in the elliptical galaxies and a globular cluster is somewhat surprising since several fairly detailed surveys of these Population II star systems have shown that they are practically devoid of eclipsing binaries and two of our theories of nova eruptions presuppose that novae are, of necessity, binary systems. Is there therefore any way by which observation and theory may be reconciled? This question is clearly a vital one as far as the binary nature of novae is concerned.

A possible answer is that although there is undoubtedly a conspicuous lack of close binaries among Population II stellar systems (and since these variables possess continuous light curves which are very characteristic, they should be fairly readily identified), we may assume that there must be some, although admittedly only a very small number. From a preliminary statistical analysis, the total number of Population II binaries we would have to postulate to account for the observed frequency of novae in these systems is of the right order. The figure comes out to be one nova every 400 to 500 years in the globular clusters as a whole; and one every thirty-five years in each elliptical galaxy.

To end the present chapter we shall consider a probable reason for this curious dearth of close binaries among the Population II stars. One of the basic features of the globular clusters and elliptical galaxies is their low angular momentum. This being the case, it can be shown that they will be composed of stars having low velocities of axial rotation with consequently very few binary systems. On the other hand, the galactic disc (Population I stars) possesses a high angular momentum and we would expect these stars to have high rates of revolution and to contain a large proportion of binary systems. It is in the galactic disc population therefore that we will find large numbers of U Geminorum variables and novae.

20. η Carinae

The main forms taken by the light curves of the novae are now fairly well known and we have seen how these may be conveniently subdivided into the three groups of fast, slow and ultraslow variables. A brief mention was made of one star which, because of its extended activity and the wide range of its light fluctuations did not fit into any of these three classes. This is the remarkable star η Carinae which will be discussed in this chapter.

As long ago as the middle of the seventeenth century it was described as varying between second and fourth magnitude and about one and a half centuries later it underwent a truly spectacular series of brightenings which has not been paralleled by any other star. Reaching magnitude 4.0 in 1810 it then climbed slowly and with many irregular fluctuations to 2.0 magnitude by about 1820, remaining essentially at this brightness for the next decade before rising to brighter than zero magnitude in 1843 when it was the equal of Sirius, the brightest star in the heavens. It then commenced a slow decline, fading to around seventh magnitude by 1870 where it has remained ever since although its light is still not steady and variations continue to occur. (Fig. 56.)

Before its observation in 1677 by Halley, we know nothing of this star apart from the fact that it is not recorded in the catalogue produced by Ptolemy which suggests, although it does not prove, that the variable was below fourth magnitude at that time (ca 140 A.D.).

The spectrum of η Carinae provides us with a great deal of information concerning this star but unfortunately spectral changes have been observed over a much shorter period than the light variations and we have no knowledge at all of the changes in the spectrum which must surely have taken place during the important period from 1810 to 1860 which includes the time when it was at its most brilliant. Undoubtedly such spectra, if they were available, would have told us much about its curious behaviour at this time. From the time when spectroscopic studies were initiated up to 1894, the star would appear to have had a predominantly absorption type spectrum, very few emission lines being visible. However, in that year, a slight fading in brightness occurred while at the same time there

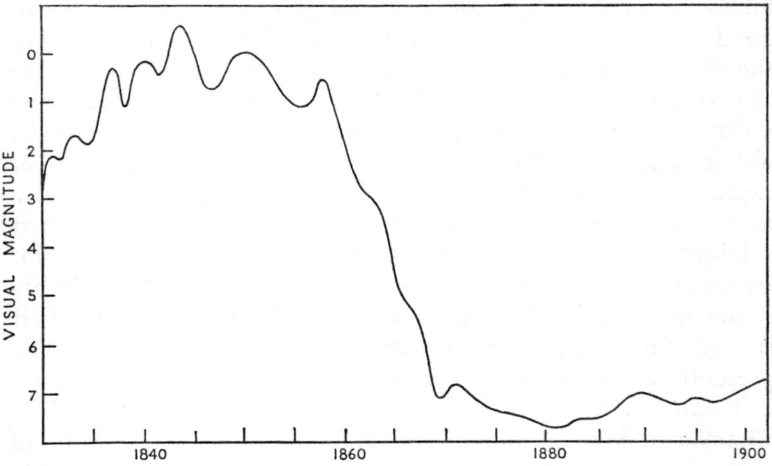

Fig. 56 – Light curve of η Carinae.

were considerable alterations in the spectrum.

Instead of an absorption type spectrum it now became one consisting in the main of emission lines and this has been the case ever since. Much more recently, the infra-red spectrum of this variable has been investigated, notably by Thackeray who has identified the strong, sharp lines of the Paschen series of hydrogen together with those due to neutral nitrogen, doubly ionised argon and possibly singly ionised titanium. The lines of other metals have also been found as well as many, including several of the strongest lines in this region of the spectrum, which have still resisted identification.

Photographs taken of η Carinae clearly show the presence of faint condensations very close to the variable which have been attributed to a large nebulous shell that surrounds the central star. From the spectrum which incidentally shows certain features very like those of the P Cygni stars, a velocity of approach of about 480 kilometres per second has been derived for this expanding envelope.

The absolute visual magnitude may, of course, be estimated if we can determine the distance of η Carinae. This may be found in several ways, all of which give values which are in good agreement. It will be recalled that one method of finding the distance of Nova Aquilae (1918) was by means of the rate of expansion of the gaseous shell surrounding this nova. Since η Carinae is also surrounded by such a shell we may use it to obtain a figure for the distance from a knowledge of the rate of expansion and its present angular diameter. Another way is by measuring the intensity of the interstellar lines of ionised calcium in the spectrum, although here we must make certain, reasonably valid, assumptions about the density of this interstellar matter lying in this particular direction. A third method is from the spectroscopic parallaxes of the nearby stars. The presently accepted distance is about 6,400 light years.

Taking into account the effect of interstellar absorption, Thackeray has estimated an absolute visual magnitude of around −14 at the time during 1843 when the variable attained its maximum brightness of 0.8 magnitude. For several years after this event, the star must have remained at an absolute visual magnitude close to −13, beginning to fade only about 1860 to something like −6 where it has stayed for more than a century. Visual observations prior to 1830 – and some of the very early ones are naturally somewhat uncertain – indicate an absolute magnitude varying between −9 to −11 for the star between 1677 when it was first mentioned by Halley as being of the fourth magnitude and 1820. It is far from easy to visualise the stupendous outpouring of energy from this star over the total period of observation; certainly it must be equal to, or more than, that of any other known star.

Even now it is extremely difficult to assign η Carinae to any particular class of variable. Its singular light variations alone put it into a class by itself. The light curve is so different to that of a normal nova that its inclusion in this class is little more than a matter of convenience. When we consider the light curves of the Na and Nb groups of novae, it is at once apparent that the period over which the nuclear explosion process occurs is very short indeed compared with that of η Carinae. Even supernova outbursts occur on a far shorter time scale. As has been mentioned in Chapter 19, its closest counterpart is RT Serpentis.

The great distance of this star coupled with the stupendous amount of energy which it has radiated over at least the past three centuries, tells us at once that it must be both bright and large. The brightness has already been discussed and now we must consider its mass.

We may make an estimate of mass from the mass-luminosity relationship although there are certain inherent difficulties associated with this method which inevitably lead to uncertainties in the result. For example there are several reasons for believing that η Carinae does not lie even approximately close to the main sequence and these will be discussed in detail later. In the meantime, if we make the assumption that the luminosity has not been appreciably altered during its past history, we can at least obtain an approximate figure. This works out to be about eighty solar masses with an error of perhaps plus or minus twenty per cent. This is quite a large margin of uncertainty but it does tell us that the star is very massive.

So far we have said little about the physical nature of this variable beyond the fact that it must be both large and highly luminous. From a study of the early spectra, Miss Cannon has classified the spectrum as that of a supergiant of type cF5 and our present picture of η Carinae is that of a huge central star with a surface temperature of about 7,500°K and possessing a very extensive atmospheric envelope or corona, around which has been formed a still more extensive shell produced from material which is being constantly ejected by the star.

One important point still remains to be settled. From its spectral type and brightness, the star obviously should lie far to the right of the main sequence but there is no conclusive evidence to say whether it is evolving onto the main sequence or away from it. Having such an incredibly large mass we would naturally expect such a star to run through the early stages of its evolution quite rapidly. Burbidge has suggested that here we have a star which has lost most of its available supply of hydrogen, either by nuclear reactions within the core or by ejection into the surrounding shell. Quite evidently it has now reached a point of instability but equally obviously from its light curve this instability is not such as to produce the extremely violent outbursts such as are displayed by ordinary novae. The very extended time scale of the fluctuations of η Carinae shows quite conclusively that any outbursts are on a comparatively slow and mild scale. On the basis of various evolutionary models, several astronomers have concluded that any pulsations taking place within stars with masses less than sixty-five times that of the sun will be stable. Above this limit, there will be an increasing instability in the pulsations such as are found for η Carinae. Whether this instability will eventually become so marked and result in an explosion, perhaps on a supernova scale, as suggested by Burbidge is speculation at the present time although from purely theoretical considerations it does not appear to be impossible.

We can of course look upon η Carinae as a very massive protostar which is contracting and moving towards the main sequence in the process. Again, theoretical arguments show that such stars must be either unstable and eject mass into a circumstellar envelope or turn supernova.

21. Recurrent Novae

The discussion of the novae in Chapter 19 was entirely concerned with those stars which have flared up to a maximum only once during the period over which astronomy as a science has been seriously studied. There we saw that a statistical survey of the number of observed novae in the galaxy and the fact that only the outer layers of the star are involved in the nova phenomenon prompt us to believe that certain stars may undergo the nova explosion, or something very similar to it, at more or less frequent intervals. Since the end of the last century, several of these recurrent novae have been discovered, many being thoroughly and continuously observed both visually and spectroscopically.

Naturally, a star of this sort would not be expected to have such a large amplitude as those observed for the normal novae which have periods ranging from many hundred to many thousands of years and this is indeed the case, the average amplitudes of these variables lying between six and eight magnitudes, compared with between ten and sixteen magnitudes for the novae.

One of the most celebrated of these variables is T Coronae Borealis, familiarly known as the Blaze Star. In May 1866, the star suddenly flared up from around tenth magnitude to a little brighter than second in a matter of a day or so, rivalling α Coronae Borealis, the brightest star in this small but striking constellation. Almost immediately however it began to fade and about a week later had fallen below sixth magnitude and was invisible to the naked eye. After fading almost to minimum there was a temporary revival to 7.0 magnitude but the star soon decreased in brightness once more, falling to 9.8 magnitude where it remained with slight fluctuations of the order of half a magnitude for almost eighty years. Then, in February

1946, it again rose sharply to third magnitude and repeated its earlier performance almost exactly. (Fig. 57.)

Spectroscopic examination has shown that T Coronae Borealis is a binary system consisting of an intrinsically variable bright Be star which is the seat of the nova outbursts and a red giant which is itself irregularly variable over a range of 0.7 magnitude and responsible for the much smaller light variations which are observed at minimum.

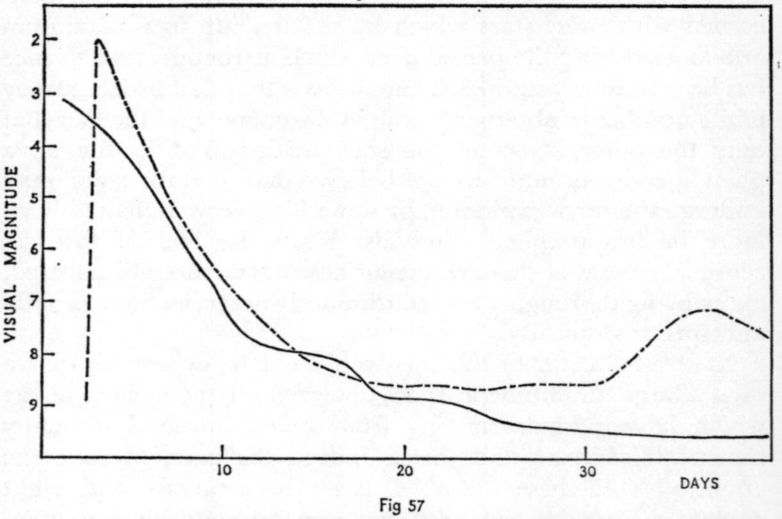

Fig. 57 – Light curves of T Coronae Borealis (*Courtesy of the British Astronomical Association*).

Other members of this class of recurrent novae have had even more frequent outbursts in recent years. For example, T Pyxidis has erupted in 1890, 1902, 1920, 1945 and 1965 while RS Ophiuchi, one of the best observed of these variables by observers in northern latitudes, has attained maximum in 1901, 1933, 1958 and 1967. It would be wrong to suppose that these stars reach the same magnitude at each outburst. The first two maximum recorded for RS Ophiuchi for instance were very unequal.

As we shall see, however, the spectral changes at each maximum are essentially the same while visually there is evidence for an expanding shell of gas around the star similar to that found around many of the novae. There is therefore very little doubt that in spite of their greater frequency and smaller amplitudes, these variables are of a similar nature to the novae and the mechanism responsible for the outbursts is in all probability the same as that which we have already discussed. We are consequently not faced with the difficulty we encountered when considering the U Geminorum stars for which there is no such direct evidence.

Data for some of the known recurrent novae are given below in Table 23.

Table 23
RECURRENT NOVAE

Star	Maximum	Minimum	Spectrum
T CrB	2.0	10.8	Q+gM3
RS Oph	5.1	11.7	Ocp
T Pyx	7.0*	14.0	Pec
N2 Sge	7.0*	16.1	—
N5 Sgr	10.1	(16.4	—
U Sco	8.8	(17.6	—

* Photographic magnitude

Light Variations of the Recurrent Novae

Like the novae, the light curves of this small class of variables are characterised by a rapid rise to maximum although generally speaking the decline is faster than that of the novae themselves, a feature we might expect since the mass of material ejected during the outburst is almost certainly less in the case of the recurrent novae. Quite often there are pronounced fluctuations during the fall to minimum and although the maxima for any particular star may differ appreciably in brightness, the shapes of the individual maxima bear a striking resemblance to each other. (Fig. 57.)

The presence of a subsidiary rise following the initial outburst, well seen in the light curves of T Coronae Borealis, is less marked in the case of RS Ophiuchi and virtually absent in those of the other recurrent novae.

Returning to the point already made that on the basis of their light curves, the recurrent novae may be associated with the U Geminorum stars on the one hand and the novae on the other, we may note that the light curves of such stars as RS Ophiuchi and T Pyxidis are indeed similar to those of the longer period U Geminorum variables, for example UV Persei and SW Ursae Majoris. The recurrent novae, having these larger amplitudes, are of course much brighter at maximum than the great majority of U Geminorum variables and one of the difficulties in making this comparison is the rather fragmentary light curves as yet available for many of the latter stars. Undoubtedly this is a problem which requires much more observational work before any definite conclusion can be reached.

Spectra of Recurrent Novae

A great deal of spectroscopic work has been carried out on the recurrent novae, some of it as in the case of the novae directed towards an investigation of a possible binary nature of these objects. Being the brightest of the class, both at maximum and minimum, T Coronae Borealis has been studied spectroscopically for several decades. Prior to the outburst of 1946, the spectrum of this variable was shown to be composite, an absorption spectrum of a gM3 star with in addition bright lines due to hydrogen, neutral and ionised helium, and doubly ionised oxygen. Following the 1946 outburst, Sanford was able to show, from long coude exposures with the 100-inch reflector at Mount Wilson, that the absorption lines possess a velocity variation which has a period of 230 days. Unfortunately, at this time, the emission spectrum contained several lines which were due to the expanding shell produced by the nova outburst and it was impossible to determine any radial velocities from this emission spectrum. Not until ten years after the eruption was Kraft able to detect velocity variations of the Hα and Hβ emission lines. A study of the absorption lines of the red giant companion indicates no abnormalities for this star which, in all probability, has filled its lobe of the equipotential surface. All of the evidence obtained from these spectra show that the blue companion is an sdBe star but more massive than a white dwarf such as those found in the systems of the old novae and possibly the U Geminorum stars. There is very little evidence at the moment for any actual eclipse in this particular binary system.

Coming now to RS Ophiuchi, Joy has obtained several spectra of this star following the 1958 outburst using a grating spectrograph on the 60-inch reflector at Mount Wilson and has been able to identify many of the observed emission lines. One of the strongest of these is that due to doubly ionised oxygen present in the outer regions of the stellar atmosphere. Of the other lines observed, a large number are those of highly ionised gases including several metals. Emission lines due to iron, for example, covered the whole range from doubly ionised iron to atoms of this metal which had been ionised no less than thirteen times!

This phenomenon of atoms in such a highly ionised state had also been noticed during the 1933 maximum of this recurrent nova and before this had been discovered only in the spectrum of the sun's corona during total eclipse. For a time it was considered that such conditions might be almost unique but since then such lines have been identified in several sources including many novae, the planetary nebulae, other recurrent novae and also in some of the symbiotic variables which will be discussed in Chapter 23.

The spectrum at minimum does show absorption lines as in the case of T Coronae Borealis, but here the similarity ends. The absorption spectrum of RS Ophiuchi appears to originate not in a red giant star but in a gaseous shell and no motion either of the absorption or emission lines has been detected which would indicate that this recurrent novae is a binary.

We must now consider one particular recurrent nova which shows so many similarities to the U Geminorum variables that it may be considered a link between the two classes. This is WZ (N2) Sagittae which underwent nova outbursts in 1913 and again in 1946. From Table 23 it will be seen that this star is a faint object at minimum and its absolute magnitude is almost certainly low, possibly in the region of +9 or +10. This, taken in conjunction with the fact that unlike the other recurrent novae which have been investigated no emission lines due to ionised helium have been observed, shows a very close correlation to the U Geminorum stars.

The absorption spectrum is consistent with the presence of a white dwarf while the emission spectrum of hydrogen shows double components. From a study at minimum light, Krzeminski has shown that this variable is an eclipsing system with a period of 81.5 minutes, the light curve at eclipse being of the W Ursae Majoris type. It will be remembered that these variables have continuously changing light curves in which the primary and secondary minima are of almost equal depth, the two components being very nearly identical and having masses similar to that of the sun. We cannot however infer that this gives us a satisfactory model for the WZ Sagittae system. In spite of the overall appearance of a W Ursae Majoris type light curve there are many anomalies to be considered. Most of the estimates of the mass of the hot companion come out too high, the smallest being 1.6 times that of the sun. (The theoretical upper limit for white dwarfs is 1.2 solar masses.) Possibly, as suggested by Burbidge, this object is a neutron star consisting of a very small core perhaps ten kilometres in diameter surrounded by a much larger envelope with dimensions comparable with that of the companion star.

Our present picture of the WZ Sagittae systems is clearly a very confusing one. The presence of a large mass of gas in the system is fairly definite. But the earlier belief that it is associated with the white dwarf companion has since been found to conflict with many of the spectroscopic observations and it appears quite probable that it may be physically detached from both stars. The inclination of the orbit is close to 90° and the eclipses are virtually total.

Masses of the Recurrent Novae Components

As we have just seen, the blue component of the WZ Sagittae system is a white dwarf or a neutron configuration and there are grave difficulties in obtaining a satisfactory estimate of its mass. If we take this to be approximately twice that of the sun, then the mass of the secondary turns out to be 0.2 solar mass.

In the case of T Coronae Borealis we are on much surer ground. From some of the spectrograms which have been obtained, mainly by Deutsch, it would appear that there is evidence for a grazing eclipse, very similar to that we have discussed in the ζ Aurigae system. The masses of the blue and red components have been estimated as 2.6 and 3.7 solar masses respectively. The blue star is therefore far too massive to be a white dwarf.

22. The Supernovae

In Chapter 19 we made the acquaintance of the normal novae which appear in fairly frequent numbers in our own galaxy and have also been observed on photographs of many of the nearer extragalactic nebulae. These are stars which usually have amplitudes of between ten and thirteen magnitudes and in which only the outermost layers of the star take part in the nova phenomenon.

But even the violence of these outbursts pales into insignificance beside the tremendous flood of energy which is emitted by the stars which we are now to discuss, stars which were originally classed with the novae and have only been differentiated from them during the last fifty years or so. The general shape of their light curves is in many ways similar to those of the ordinary novae and visually it is often far from easy to decide between these two classes of variable. Like the novae, the supernovae brighten from anonymity with an equal suddenness and subsequently decline over a period of months to something like their former faintness. There are, however, certain marked differences which clearly set them in a class apart. For example, they have amplitudes often very much in excess of thirteen magnitudes; they have spectral characteristics altogether different and spatially they are of much less frequent occurrence than the normal novae.

To introduce this chapter on the supernovae we must first mention a variable star which appeared many years before the distinction between these stars and the novae was recognised and which itself was considered to be nothing more than an ordinary nova at the time of its discovery.

In August 1885 a new star suddenly appeared very close to the centre of the spiral nebula in Andromeda, a nebula which we now know from the large number of Cepheids and long period variables discovered in it to lie at a distance of 2,200,000 light years. The brilliance of this star was such that it completely transformed the appearance of the nebula, normally just visible as a hazy patch of light to the naked eye, and for the space of a few weeks it was just visible without instrumental aid. Estimates made at the time show that it reached a magnitude slightly brighter than 6.0 about the time of maximum. It was probably first observed by Professor Gully on August 17th but he put it down to some defect in the new instrument which he was using. Similarly Max Wolf at Heidelberg observed it on August 25th and 27th but considered it was due to some optical effect caused by the bright moonlight. From the records of that time it would seem that very few astronomers, including Hartwig, realised that the changed appearance of the nebula was due to the presence of a new star. Two years after its discovery it had sunk out of sight completely, being well below sixteenth magnitude and it was universally presumed to be merely another nova, almost certainly one within our own galaxy and only seen projected against the nebula and it therefore attracted very little attention. We must also remember that at this time the nature of the Andromeda nebula and its distance were only matters of conjecture and very little was known about both.

More than a quarter of a century later in 1917, Ritchie discovered several faint novae on photographs taken of the Andromeda spiral and more than a hundred were found altogether within the following fifteen years. A study of their absolute magnitudes soon proved that at maximum they were similar to the novae which appear frequently in the galaxy and consequently, the star of August 1885 was of a very different kind altogether as far as its luminosity was concerned.

More important perhaps, during these fifteen years the controversy over the nature and distances of the nebulae was being resolved and astronomers were moving away from the view that they were fairly close objects and coming to accept that instead they were island universes, very similar to our own, at inconceivably great distances. With the final proof of their true distances from the period-luminosity relationship for the various classes of variable stars found in them, they attained their true status and once this was realised, it threw a new light on the nature of the 'nova' of 1885.

If we proceed on the assumption that this star was indeed within the boundaries of the nebula and not simply a foreground star inside the galaxy, then the fact that its estimated brightness at maximum was 5.5 magnitude means that its luminosity must have been comparable with the integrated brightness of the whole nebula, equivalent to 100,000 million suns and secondly its absolute magnitude must have been

Table 24
BRIGHT EXTRAGALACTIC SUPERNOVAE

Nebula	Photographic maximum magnitude Apparent	Absolute
NGC 224	6.4	−15
NGC 1003	12.8	−14
NGC 4273	14.4	−12.3
NGC 4303	12.8	−13.8
NGC 4321	11.9	−14.8
NGC 4424	11.1	−15.6
NGC 4486	12.0	−14.7
NGC 4527	13.0	−13.7
NGC 4621	12.0	−13.0
NGC 4725	12.8	−13.6
NGC 5253	8.0	—
NGC 6946	12.9	−12.4
IC 4182	8.2	−16.6

about −17 which is far greater than that of any previously known kind of star.

Once this astounding fact was recognised it naturally prompted an intensive photographic search of other extragalactic nebulae which soon resulted in the discovery of other supernovae of this kind and a serious study of these highly unusual objects has provided us with a much fuller explanation of their unique light and spectral features.

The following table give a list of some of the brightest supernovae and also the nebulae in which they appeared.

It can be seen at once from the absolute magnitudes given in the above table that in general these stars are in quite a different class to the normal novae and indicate a luminosity at maximum more than 100,000,000 times that of the sun. To illustrate this more dramatically, a single supernova during the time it is visible, often only a few months, emits as much radiation as the sun does in several million years. If a supernova such as that which appeared near the nucleus of the spiral nebula in Andromeda were to suddently flare up in our own galaxy at a distance of even 350 light years from the sun it would appear as bright as the full moon but would still only look like a star without any visible disc even in the largest telescopes.

The output of energy associated with even a single supernova outburst is utterly beyond anything else known in the universe with the exception of the recently discovered quasars. Were the sun to turn supernova – later we shall see that it is extremely doubtful if this could occur – the result would indeed be catastrophic. All of the planets, without exception, would be vapourised within hours. Fortunately for us, recent theoretical calculations have shown the unlikelihood of this, the reason having been briefly mentioned earlier in the chapter on RR Lyrae variables. We saw there that provided the mass of the helium core of these stars does not exceed 1.44 times the mass of the sun (this critical figure being known as Chandrasakhar's Limit), they will follow a well-defined path of evolution once they begin to evolve away from the main sequence ending as white or red dwarfs. Later in the present chapter we shall see that where the mass of the helium core does exceed this limit we have the necessary conditions for the star to experience a supernova outburst at some time during its lifetime. But first we must consider the light variations of the supernovae and the typical spectral changes which accompany them.

Light and Spectral Variations in the Supernovae

In the case of these stars it is convenient to consider both light curves and spectral changes together. A comparison of the light curves of the supernovae with those of the novae show that the former are much less diverse and from those studied in detail it has been found that the light and spectroscopic changes keep so well in step that even if a supernova is not discovered until after maximum is past we can generally tell simply from an examination of its spectrum during decline just when light maximum occurred. In spite of the great similarity of the light curves of these stars, Baade has classified them into two types, basing his classification on their spectra, the general shape of the light curve at maximum, and their absolute magnitudes.

Baade's Type I supernovae have the higher absolute magnitudes lying between −14 and −17 and have light curves which are characterised by a sharp maximum which is succeeded by a fairly rapid decline that eventually slows until the star has faded to minimum. S Andromedae, the supernova of 1885 and that which appeared in the nebula IC 4182 were clearly of this type. The spectra taken after the explosion show that the expanding cloud of gas is very deficient in hydrogen and it is generally agreed that these stars are of Population II, very old stars, formed from the primeval matter of the galaxies. At maximum they appear reddish but the colour gradually fades until the colour index becomes about +0.5. The mass of the envelope thrown off by the explosion is almost invariably less than that in the case of the Type II supernovae being of the order of 0.1 solar mass and in addition, the radial velocities of the ejected gases are lower, usually about 1,500 to 2,000 kilometres per second.

Baade's Type II supernovae on the other hand remain around maximum for a longer period than stars of the former class and are decidedly blue at maximum brightness, implying that in their case there is a large bolometric correction to be made. These supernovae have so far been found only in the spiral arms of the galaxies and obviously belong to the Population I stars. By contrast to the Type I supernovae, the expanding envelopes are found to be relatively rich in hydrogen and the outward moving gases have velocities up to 5,000 kilometres per second. The mass of this gaseous envelope is also much greater being between five and ten solar masses. Their absolute magnitudes generally lie between −12 and −13.5 and in their spectra are found certain similarities to those of the ordinary novae which has led some astronomers to suggest that they may form an intermediate link between the novae and the Type I supernovae. Whether this is actually the case is still much a matter of conjecture but certainly a lot of the evidence points in this direction.

Masses of the Supernovae

Observation has shown that the two types of supernovae possess masses of a different order. Those of Type I, the older Population II stars, have masses not very different from the sun which explains why the mass of the ejected gases is relatively low. The Type II supernovae however have much larger masses, almost certainly as high as thirty times that of the sun, possibly higher, and we may assume that these stars are, in all probability, still associated with the parent gas clouds from which they have condensed.

When considering the reactions which are likely to lead to a supernova explosion we shall find that the existence of these two extremes of mass is very important since it leads us to two different nuclear processes which are thought to take place in these stars prior to the supernova detonation.

Frequency of Supernovae

Before going on to look at the various nuclear reactions going on inside these stars we must first examine the frequency of supernovae of both types in the various galaxies including our own. The photographic survey carried out once the reality of these unimaginably violent detonations was recognised resulted in the discovery of a large number of such stars and at first sight it might seem that they are of fairly frequent occurrence. We must bear in mind however that the survey covered several thousand galaxies and we now know that the supernovae are far rarer than ordinary novae appearing, on average, only about twice a millenium in any one galaxy. This does not imply that the interval between successive supernovae in any particular galaxy is necessarily long. There are instances known where two have appeared in the same galaxy within a very short time of each other. For example, two such stars flared up in NGC 4321 in 1901 and again in 1914 while in our own galaxy, of the three supernovae which have been identified with certainty, two appeared within thirty-two years of each other.

If the star which erupted in the constellation of Puppis in 1006 was a supernova, as seems to be the case from the records which are available, then it shone out only forty-eight years before the supernova of 1054.

The exceptionally brilliant star discovered and studied by Tycho Brahe in 1572 in Cassiopeia was clearly, from the details we have of its light curve (Fig. 58), a Type I supernova. Since the heavens were at that time regarded as unchanging and unchangeable, this object was generally regarded with mixed

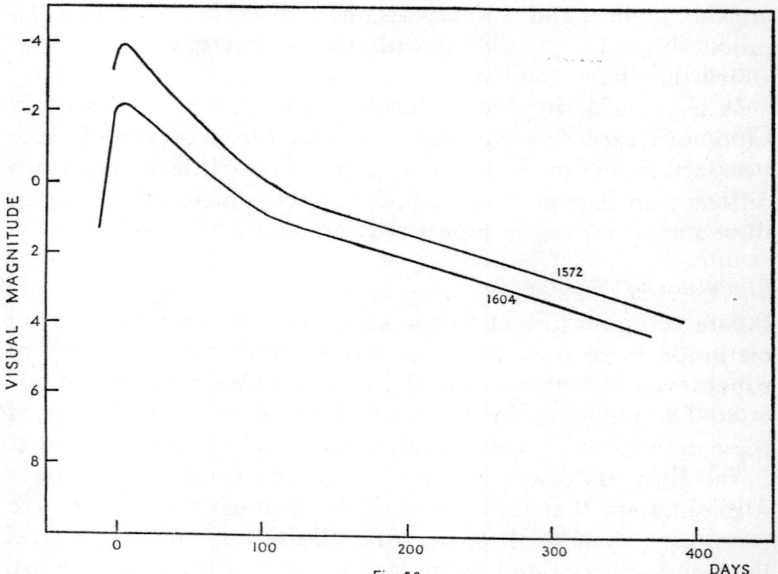

Fig. 58 – Light curves of the supernovae of 1572 and 1604.

awe and superstition. Although it appeared some years before Tycho's observatory was built and therefore we know its position only with a certain degree of dubiety, recent systematic searches of the region have failed to reveal it beyond all doubt and it is undoubtedly fainter than magnitude 18.0. Since it outshone Venus at maximum, being visible in broad daylight, it must have had an amplitude of more than twenty-one magnitudes; a very remarkable object indeed!

Similarly, the object discovered in Ophiuchus in 1604 and closely followed by Kepler also possessed a light curve very like that of a Type I supernova. (Fig. 58.) Had it been a normal nova its present magnitude could not be much less than 12.0 since the amplitudes of ordinary novae do not exceed fourteen magnitudes. Long exposure photographs taken in red light of the area have revealed a very small nebulosity close to the stated position which may possibly be the remains of the star or more probably the gaseous shell which was thrown out by the explosion, but there is certainly no sign of the star itself.

The third galactic supernova of which we have any detailed knowledge appeared in Taurus in the year 1054, the appearance of a bright new 'guest' star being chronicled by ancient Chinese and Japanese astronomers. At its maximum it was the equal of Jupiter, magnitude −1.9, and in its present position is the mysterious and beautiful Crab Nebula, one of the strangest of all the nebulae. Spectroscopic and photographic examination of this nebula show that it is evidently still in the process of expansion. The present measurements show that it is expanding at a rate of 800 kilometres per second and by extrapolating backward in time we find that the detonation which brought about this expansion must have occurred around 1054.

The Crab Nebula – a Supernova Remnant

The distance to the Crab Nebula has been estimated by several methods at almost 5,000 light years which means that the star actually exploded more than a thousand years before the Pyramids were built on Earth. This tremendous distance also means that if, as recorded, its magnitude at maximum was −1.9, its absolute magnitude must have been a little more than −13. Its luminosity at maximum has been calculated as being the equivalent of 300 million suns placing it quite definitely in the Type I supernova category.

The Crab Nebula itself shows both a continuum and several bright emission lines in its spectrum and when photographed in red light, the outer regions are found to have a curiously filamentous structure which is the source of the bright line spectrum. The central portion on the other hand is an irregular mass in which little actual structure can be observed and it is here that the continuous spectrum originates. Since the hydrogen emission lines are abnormally faint it may be assumed that this element is present only in low abundance, a condition which we have seen applies to the gaseous remnants of all Type I supernovae. We therefore have here an independent confirmation of its nature.

The luminosity of the nebulous gases has been found to be due to one of two very faint stars situated close to the centre. As yet there is insufficient evidence from observation to say with certainty which star is the remnant of the supernova but whichever it is, Minkowski has estimated that it must have a surface temperature somewhere in the region of 500,000°K and a luminosity 30,000 times that of the sun but with a radius only one-fiftieth of the sun. The total amount of matter in the nebula has been reckoned by some observers as close on fifteen solar masses (a figure which, if true, would seem to place the supernova in the Type II class). The stellar remnant of the 1054 supernova is clearly still evolving towards the white dwarf stage even now judging by the high luminosity and temperature of the nebular gases.

One further important feature of the Crab Nebula deserves mention here, namely that it turns out to be an intense emitter of radio waves. Many quite large radio sources are known within the galaxy and it has been shown that this emission may be explained by supposing that here we have vast clouds of interstellar gas which absorb the light from intensely hot main sequence stars embedded within them. During this absorption process, the hydrogen atoms which make up the gas clouds are ionised, i.e. broken up into free protons and electrons which are then able to move independently through the cloud. Inevitably, there are collisions between these particles and when this happens they may either recombine or simply pass sufficiently close to each other for certain physical interactions to take place without recombination to form neutral hydrogen atoms.

In the former case, visible light is emitted accounting for the brightness of the clouds. In the latter instance, the emitted radiation is not confined to the visible region of the spectrum. Ultra-violet and infra-red radiation is emitted together with radio waves. The Orion Nebula is one such vast cloud of gas which has been distinguished as a radio source and many more are known.

However, when we come to take a closer look at the Crab Nebula we find that the radio emission is far more powerful than that received from the huge gas clouds and it is no longer possible to explain this intense radiation in terms of simple particle collision processes. We must search for a deeper cause and the clue to this is provided by the tremendously high velocities of the gaseous filaments in the outer environs of the nebula, velocities imparted to these gases by the original supernova explosion. Although the nebula is expanding at the rate of about 800 kilometres per second, within this filamentous structure there are certainly far higher velocities present of several thousand kilometres per second.

Here we must introduce the concept of interstellar magnetic fields since these play a very important part in the physical reactions taking place. Although the general magnetic field existing among the stars is extremely small compared with that around the terrestrial poles, the intensity is appreciably increased in a situation where filaments of gas are in violent motion and collision.

Alfven and Herlofson have put forward the idea that under conditions such as those prevailing inside the Crab Nebula, the electrons are accelerated until their energies approach those of low-energy cosmic rays. Once this occurs, radio waves are emitted as these high speed electrons move through the superimposed magnetic fields.

Spectra of Supernovae

We might think from a study of the light curves of supernovae that, in spite of the fact that the outburst is obviously on a far greater scale than the novae, their spectra should, at least, have a great deal in common. This, however, is not what is found in practice. When we examine the spectrum of a supernova, we find that it is totally unlike that of any other known kind of star. Indeed, the differences are so great that many aspects of these spectra have still not been unravelled. From what is already known, it would seem that many of the observed lines are due to hydrogen and iron which are in the form of a highly incandescent gas being hurled outward from a central point of detonation at extremely high velocities sometimes, though rarely, as high as 8,000 kilometres per second. Clearly we are here dealing with something on a vastly different scale to the explosions produced by the ordinary novae. We are now in a position to go on to consider the physical and nuclear reactions which can take place with a certain type of star to lead to the inconceivably incredible explosion of a supernova outburst.

The Supernova Explosion

It is immediately obvious from the abruptness of the outburst and the colossal amount of energy liberated in so short a space of time that the reactions responsible must be of a truly catastrophic nature. From the amount of material ejected in the course of the outburst we clearly cannot fall back upon the explanation we gave for the novae since far more than merely the outer skin of the star is involved in the process. To seek the underlying cause of the explosion we must look deeper into the star and here we go back to the important point made in Chapter 5 concerning Chandrasekhar's Limit. Provided that the core of helium in the centre of the star does not have a mass exceeding 1.44 times that of the sun the star is able to evolve quite smoothly to the right of the main sequence since the degeneracy pressure in the interior can balance the gravitational pressure as the star contracts in its evolution towards the final dwarf stage.

In the case where the mass of helium does exceed this limit, the extremely delicate balance cannot be maintained. Degeneracy pressure, the crushing of the material within the core bringing about virtually complete collapse of the electron structure around the nucleus, is simply not powerful enough to balance the pressure of the overlying layers of gas and as a result the core cannot expand and the temperature must continue to rise, this being the only other alternative. For part of its lifetime, nuclear reactions are sufficient to maintain this temperature balance but as time goes on a limit is reached beyond which nuclear reactions fail to provide the necessary energy and the one alternative left is gravitational shrinkage of the core.

We have already seen that with the formation of a helium core in the centre of the star little, if any, mixing of the hydrogen and helium occurs and a relatively thin skin is formed in which hydrogen burning takes place, the helium produced going to swell the core as the star evolves.

A time comes however where the gravitational contraction of the core raises the internal temperature sufficiently for the helium, normally a very inert material, to begin burning with the formation of still heavier elements. In particular carbon, oxygen and neon are formed. Certain subsidiary reactions have also been shown to occur at this stage due to a very small amount of mixing of hydrogen with helium in the burning zone producing most of the known chemical elements and free neutrons. The structure of such a star at this stage is almost identical with that where the mass of the helium core is below Chandrasekhar's Limit, the only basic difference between them being that a far larger quantity of hydrogen is present in the outer layers of the star.

However, as the evolution of these more massive stars proceeds, the core now consisting of oxygen and neon approaches Chandrasekhar's Limit and at this stage the degeneracy pressure within the core is no longer able to maintain the critical balance. Consequently, the core shrinks still further, the temperature climbs still higher to somewhere in the region of 600 million degrees when first neon and then oxygen begin to burn producing what is known as the silicon group of elements.

This is not however the end of the story. There is still one further stage in the life history of such a star before the supernova explosion. In the centre, the core continues to contract pushing the internal temperature up still further until at around 4,000 million degrees the silicon group of elements begin to burn forming the metals of which iron is the most abundant.

An important point to realise about these particular nuclear reactions is that one of the most dominant processes, perhaps *the* most dominant process, is the transformation of protons into neutrons. In Chapter 5 we saw that when protons are converted into neutrons, particles known as neutrinos are emitted which have the peculiar property of taking part in no known nuclear reaction but simply of carrying energy away from the star and out into space. At these incredibly high temperatures the loss of energy due to the emission of these neutrinos is extremely rapid and efficient and the star is forced to continue shrinking at an ever-increasing rate to make good this large energy loss. By the time the iron group of elements begins to react the internal temperature has reached the staggering figure of 5,000 million degrees but now, instead of building up still heavier elements, the result is complete conversion back into helium!

At first sight this would appear to be not only completely unexpected, but utterly inconceivable. Nevertheless, this fact is based upon well-known principles and the explanation is really quite simple and amply confirmed by theory. All of the nuclear reactions we have been considering so far are known as *fusion* reactions since the heavier elements have been gradually built up by the fusion of lighter nuclei. Now nuclear reactions can only go on even at these enormously high temperatures if they lead to a release of energy which happens only when nuclei with a low binding energy per neutron (or proton) produce nuclei with a higher binding energy. It has been demonstrated, however, that the nuclei which are heavier than iron have lower binding energies than iron itself (such elements can produce energy by the opposite process of fission). The next step after the production of iron must therefore be a *fission* reaction and the one which occurs is that which transforms iron back into helium.

The net result of this is that if not all, certainly a very high proportion of the energy produced during the nucleosynthesis of the various elements throughout the lifetime of the star is now absorbed in this final stage. Gravitational infall occurs on a catastrophic scale and in an extremely short space of time, possibly within ten seconds.

It is not this internal collapse however, cataclysmic as it is, which produces the supernova explosion; rather it is the inescapable fact that in collapsing, the temperature of the outer layers where the earlier nuclear reactions are still going on, such as helium, oxygen, neon and hydrogen burning, increases rapidly and it is this abrupt increase in temperature of these highly reactive species to somewhere in the neighbourhood of 300 million degrees which is the ultimate cause of the detonation. We have already seen that during a normal nova outburst only the outermost layer of the star is affected. This is also the case with a supernova but here the amount of material is far greater, about ten thousand times more than is ejected during the nova phenomenon.

Once the star has lost so much of its mass, then the stage is set for it to continue to evolve fairly smoothly to the white dwarf stage since the ejection of this material now brings its mass well below Chandrasekhar's Limit.

In spite of the much greater violence of a supernova explosion, the velocities with which the gases are expelled from the star are not appreciably greater than those of the novae. This would indicate that the energy which is released by both processes is almost identical. It is only the great difference in the mass of the gas which in the main gives rise to the observed differences between the two types of outburst.

The preceding discussion is, of course, a somewhat simplified idea of the various reactions going on prior to a supernova explosion. To complete the discussion, however, we must look just a little further into the situation since as we have seen there are two types of supernovae which differ considerably in their masses and it would seem quite probable that whereas the above sequence of nuclear reactions may very well explain one type, it cannot feasibly explain both. It is certainly necessary that whatever the mass of the star it must exceed the critical limit which has just been outlined. Now we must emphasise here that the star must not just exceed this mass when it is born out of the parent pre-stellar cloud, it must also do so when the internal regions have reached an advanced evolutionary state. This is particularly so for the Type I supernovae which have the smaller masses. There will naturally be some loss of mass during the initial stages of evolution. For the Type I supernovae this will be negligible since they have fairly small masses to begin with; for the Type II supernovae with masses between twenty-five and thirty times that of the sun, we might expect considerable mass loss. However we must here take into consideration the fact that such supergiant stars evolve extremely quickly and the very high internal temperature we have postulated will be attained long before these stars have moved very far from the main sequence, so here again the mass loss should be quite small by the time the conditions are suitable for the star to become a supernova.

We can now see that the sequence of reactions we have just suggested are those most likely to be associated with the Type II supernovae, i.e. those which show a high proportion of hydrogen in the expanding envelopes. For such stars, however, recent calculations have shown that the material in the centres may be non-degenerate and the implosion of the core is the basic cause of them turning supernova.

Having made this point let us now turn to a consideration of the Type I supernovae, those with small masses where core implosion is not the cause of the outburst. Here it is assumed that virtually all of the star is composed of degenerate matter prior to the explosion, this taking place when the central temperature reaches about 3,500 million degrees. Before this happens, the internal temperature can reach as high as 2,000 million degrees which is above the explosion temperature of the nuclear fuels in the interior but no explosion occurs since the material must first become degenerate. It is when these two conditions are satisfied, a sufficiently high temperature and about ninety per cent of the star composed of degenerate matter that detonation occurs. Since by this time the energy-producing skin will have virtually reached the stellar surface, we would expect there to be very little unused hydrogen remaining in the outer layers and consequently very little in the expanding shell after the outburst, a fact which is confirmed by observation of this type of supernova.

Of course, not all astronomers subscribe to these particular theories. Anything we know for certain about the origin of supernovae is pitifully small compared with the multiplicity of questions which remain to be answered. From what has already been said, however, it is apparent that very complex reactions take place during a type I supernova explosion all within a few seconds. The fundamental characteristics which have emerged from a study of these processes are that whereas the Type I supernovae depend upon a large proportion of the stellar material being degenerate, this is not necessarily so with the much more massive Type II supernovae where we need postulate nuclear fusion only up to the point where the iron group of elements is formed.

At first, it was considered that since hydrogen burning will contribute energy to the exploding system, the Type II supernovae would be the brighter of the two at maximum whereas we know from visual observations that this is not the case. There are several reasons for this. We must remember that the maxima of Type II supernovae are broader than those of Type I and the integrated area under the light curve is greater for the former than for the latter. Also, the light curve itself is giving us only a measure of the amount of visible radiation emitted and there is now growing evidence of very strong emission in the ultra-violet from a Type II supernova. We saw earlier that these stars are pronouncedly blue at maximum and that large bolometric corrections have to be applied in their case.

Recently an alternative theory has been put forward to explain the phenomena of a Type I supernova detonation. Since the light curves of these stars appear to decay exponentially after the initial rise to maximum it has been suggested by Burbidge, Burbidge, Fowler and Hoyle that here the cause of the explosion is not fusion of light nuclei (as in Type II supernovae) but rather the fission of much heavier elements, in particular Californium-254 but also several of the other radioactive transuranic elements, these being spontaneously fissionable and possessing half-lives of from a few minutes to about one year.

Since the structure of a Type I supernova, before explosion, is nearly all degenerate, the material of which it is composed is highly unstable and it can be shown theoretically that because this is the case, the explosion, when it occurs, is sufficient to disrupt the entire star and with almost all of the energy being concentrated in the centre, it produces a shock wave of tremendous violence which passes outward through the helium zone in about ten seconds. In so doing, it causes the helium nuclei which have already been built up into carbon to add on further alpha particles (helium nuclei) to form elements of a much greater atomic weight and these are then capable of acting as 'seed' nuclei for the formation of the radioactive transuranic elements.

Clearly we must await the accumulation of far more comprehensive observations before any final conclusion can be reached on the origins of supernovae. A vast amount of observational and theoretical work has already been carried out but unlike most other types of variable star, it is only recently that these objects have been subjected to an adequate study.

Supernova Remnants

To end the present chapter we must consider the end product of a supernova explosion, namely the gaseous remnants left once the detonation has occurred. One of these, the Crab Nebula, has already been discussed in detail and this is unique in that we also have visual evidence of the supernova itself. It has also provided us with vital information which enables us to state, with a fair degree of certainty, whether a particular gaseous nebulosity is likely to be a supernova remnant or simply a cloud of gas like the nebula in Orion.

Two of the best observed supernova remnants within the galaxy are found in the constellations of Cygnus and Cassiopeia, designated Cygnus Loop and Cassiopeia A respectively. Both are exceptionally intense radio sources, being even stronger than the Crab Nebula, and photographs of these regions have shown the presence of large numbers of filaments which are in rapid motion. A piece of evidence which provides additional confirmation that these gaseous filaments are the product of a supernova explosion is that the intense radio emission from these sources comes, not only from the filaments themselves, but also from regions where these cannot be seen optically.

Cassiopeia A has been estimated to be 11,200 light years distant and to have a diameter of just over thirteen light years. The Cygnus Loop is much nearer being only about 2,500 light years away but with a diameter ten times as great. Calculation has shown that the mass ejected by the Cassiopeia supernova was similar to that of the sun whereas that ejected by the supernova which formed the Cygnus Loop must have been almost six solar masses. The approximate ages of these supernova remnants have been given as 260 and 50,000 years respectively. Unlike the Crab Nebula, both of these radio sources are the remnants of Type II supernovae.

The fact that supernova remnants have been found to be powerful transmitters of radio waves suggests that such objects should be capable of detection in other galaxies. There is however a grave difficulty here. A large number of radio sources have been discovered in recent years outside the galaxy, most of these far more powerful than the supernova remnants we have just been discussing, but in this case the radio emission is due to the collision of two galaxies, something on a far vaster scale than the galactic radio sources. Quite clearly, it is impossible to show the existence of supernova remnants outside the galaxy unless we can pinpoint small, discrete radio sources within the framework of the galaxy as a whole. This will only be possible with the nearer stellar systems which possess large apparent angular diameters. We must therefore look to the Magellanic Clouds for evidence of extragalactic supernova remnants.

So far, only three emission nebulae situated within the Large Magellanic Cloud have been identified from their optical and radio properties as remnants of supernova explosions, and from a comparison with similar objects in the galaxy it has been shown that all three were produced by Type II supernovae. Now this is quite an interesting observation since previous surveys of irregular nebulae (such as the Magellanic Clouds) have revealed only Type I supernovae.

Since we already know the distance of the Large Magellanic Cloud to be about 180,000 light years, we are in a position to calculate such vital parameters as the diameters and the total energy radiated by these objects with a good degree of precision and it is gratifying to find that they agree very closely with those calculated for galactic supernova remnants of Type II.

One further feature of the Large Magellanic Cloud deserves mention here; a curious phenomenon known as Constellation III which is clearly visible in photographs of the region taken in ultra-violet light. This huge arc formed of high luminosity blue stars was named by Shapley and it appears that it is, in reality, part of a vast shell, the inner regions of which contain little, if any, neutral hydrogen. Of the three Type II supernova remnants found in the Large Magellanic Cloud, two are embedded within one edge of this shell.

Constellation III bears such a striking resemblance to an actually expanding shell that it is tempting to explain it in terms of some gigantic explosion which must clearly have occurred at some time during the early formation of the galaxy. It is true that such an explosion must have been on a vastly larger scale even than that of a supernova and the dimensions and mass of the exploding 'star' almost beyond comprehension.

Nevertheless, such super-supernova outbursts have been suggested by Shklovsky to explain the formation of the heavy elements within galaxies. During the very early stages in the evolution of a galaxy, the gases will consist almost entirely of hydrogen and although reactions can be postulated to explain the formation of the lighter elements, very serious difficulties arise when we try to explain the presence of the heavier elements. It simply appears that there is no known sequence of reactions which will eventually lead to the production of these elements *except those taking place in the centres of certain types of stars*.

Theoretical calculations have indicated that these stars will be the Type I supernovae and here, too, we have the necessary mechanism for scattering these elements throughout the mass of galactic gas. On Shklovsky's hypothesis, the Type I supernovae which occur in the early evolutionary stages are far more powerful and frequent than during the later life history of a galaxy since these extremely massive stars, which are among the first to be formed out of the gas cloud, contain tremendous reserves of hydrogen. Now we can make an estimate of the mass of Constellation III and hence that of the material ejected during this explosion. As shown by Westerlund and Mathewson this is of the order of 100,000 solar masses; a truly stupendous figure! Once the ejected material slows sufficiently, it will then be possible for star formation to occur and it will be from among these stars of spectral classes O and B that the Type II supernovae whose remnants have been observed will originate.

Two other similar spherical shells have been found in the Large Magellanic Cloud and several have been observed in other extragalactic nebulae. Some additional evidence in favour of Shklovsky's hypothesis of super-supernovae explosions is provided by the discovery of Lynds and Sandage of a gigantic explosion in Messier 82. Here we may indeed be witnessing one of the most grandiose phenomena in the whole of the universe.

23. Nova-Like Variables

In the last four chapters we have discussed the novae, recurrent novae and supernovae; all stars which undergo explosive eruptions of a varying degree of violence. It is now our purpose to consider some classes of variable stars which at times show characteristics somewhat similar to the eruptive stars either in their light variations or spectral features but which cannot, by any stretch of the imagination, be fitted satisfactorily into any of the preceding groups. Without doubt, this is one of the most heterogeneous classes of all, containing a wide diversity of stars, many of the individual types bearing only a superficial resemblance to each other. Here it is quite certain that finality has not been reached and perhaps never will be for some groups contain such dissimilar stars that further assessment and amendment of classification is inevitable.

Symbiotic Variables

Remembering the enormous difference in size, density and temperature to be found among the stars, we might expect to come across some very unusual stellar associations among the variable stars, indeed we have already encountered such cosmic curiosities among the eclipsing binaries with stars such as VV Cephei where the two components may be said to represent the extremes in both size and temperature. The diverse nature of the two components of such systems is nowhere so marked as among the symbiotic variables, named by Merrill and extensively investigated by him. Like the nebular variables discussed in Chapter 16 this is a somewhat mixed group of stars although all are in some way intimately connected with nebulosity.

A feature common to all of these variables is their combination spectra consisting of bright emission lines (very like those which are so characteristic of the gaseous nebulae) which are superimposed upon a low temperature continuum suggestive of the presence of a red giant star of a similar nature to the long period variables. Indeed, one of the stars generally included in this class (and which has been discussed elsewhere) is R Aquarii, usually classified as a Mira type variable with a mean period of 386.86 days and an amplitude of about five magnitudes. The presence of the bright nebulosity surrounding this star is shown clearly on long exposure photographs and there is little doubt

Fig. 59 – Light curve of Z Andromedae.

that this nebula contributes to both the light and spectroscopic peculiarities of the variable. There is also evidence for a hot, early type companion immersed in the nebulosity, a combination considered to be a characteristic feature of most of these variables.

One of the first stars to be placed in this class and also one of the best observed is Z Andromedae. The light curve of this particular variable is shown in Fig. 59 from which it will be seen that although in general the light changes appear to be irregular there does seem to be an underlying period of about 300 days and evidence for another of 700 days. It is quite probable too there may even be an overlapping, much longer period of close on thirty years. The amplitudes of the two longer periods is of the order of 1.5 magnitudes. Even taking all of this into account, we have not exhausted all of the peculiarities of this star's light curve for in 1901, 1914, 1939 and again in 1956 there were more spectacular outbursts, very reminiscent of

Table 25
SYMBIOTIC VARIABLES

Star	Maximum	Minimum	Spectrum
Z And	8.9*	12.4	Pec+Mo
R Aqr	6.7*	11.6	M7e+P+O
CM Aql	14.4*	16.5	Ocp
GV Aql**	12.5*	16.8	—
AE Ara	11.6*	13.8	Beq
Z CMa	8.9*	11.4	Beq
AG Car	7.2*	8.5	Beq
DI Car	11.1*	13.0	Pec
BF Cyg	9.3*	13.4	Bep+gM4
CI Cyg	10.8*	13.4	Pec+gM4
RW Hya	9.7*	10.9	K5ep+M2ep
SY Mus**	11.3*	12.3	Pec
AG Peg	6.4	8.2	Bep
AX Per	10.8*	12.5	Pec
RX Pup	11.1*	14.1	Pec
V Sge	9.5	13.9	Ob
SS Sge	12.0*	16.8	—
FN Sgr	9.0*	14.2	—
HK Sgr	13.1*	15.8	Pec
CL Sco	11.2*	13.9	Pec

* Photographic magnitude
** Probable member of the class

nova eruptions, when the star attained ninth magnitude.

Quite clearly we are here dealing with objects which are not easy to classify. It would appear from the researches which have already been carried out on these stars that in order to differentiate them from other close binary systems which are also subject to explosive outbursts such as the U Geminorum variables and the novae proper, we must now define certain conditions which must be fulfilled before any particular variable can be assigned to this class otherwise the situation will become hopelessly confused.

These conditions are that the symbiotic variables must be binary systems consisting of a hot dwarf star and a red giant or supergiant companion which fills its lobe of the first equipotential surface and the orbital period must be in excess of 100 days. This latter condition is added to exclude such variables as the old novae which have orbital periods of only a few hours.

A list of some of the symbiotic variables is given in Table 25 but it must be remembered that several of these stars have been poorly observed, particularly the fainter members and future research may well result in the transfer of certain stars from this class to another.

Spectra of the Symbiotic Variables

As mentioned earlier, high-excitation emission lines are visible in the spectra of these stars, mainly of neutral hydrogen and singly ionised helium, silicon and iron, together with certain forbidden lines of doubly ionised neon, iron and oxygen. At certain phases, too, lines of even more highly ionised elements are present, for example those of iron ionised no less than nine times!

The underlying low-temperature continuum of a cool, red star from which a surface temperature of between 4,800°K and 5,300°K has been calculated for these companions exhibits the characteristic absorption bands of titanium oxide at certain periods and the combination of these two extreme types of spectra allows us to make some assumptions concerning the nature of these extremely peculiar variables.

The original suggestion that here we have an association of a red giant very similar to the long period variables and a much hotter *comes* provides the simplest explanation and has much to commend it. Among the long period variables and some of the recurrent novae, we have similar pairs of stars, e.g. Mira itself and T Coronae Borealis but in the case of the true symbiotic variables there appears to be an added complication, namely an extremely large nebulous envelope which, in all probability, encompasses both components. Attempts have been made to establish the density of the material of this shell of gas but all that can be said at present is that it seems to be intermediate between that found in the planetary nebulae and that of the disc of gas surrounding shell stars such as Pleione in the Pleiades.

Owing to the highly irregular nature of the light changes, it is not possible to say whether there is any real evidence for eclipse although since more than a score of these variables are known it appears likely that some of them should be so orientated in space that the plane of the orbit lies very close to the line of sight as seen from Earth. The same difficulty arises when we try to obtain orbital velocities from the lines in the spectrum. The variations are usually so complex that in many instances we must assume that these come from turbulences within the system and the task of separating orbital velocities from these other effects is almost insurmountable.

More recently some astronomers have advanced the theory that these peculiar variables are not binary systems at all, that they are merely single stars which are totally enveloped by a cool gaseous shell which has a far higher density than the extended chromospheres of normal giant or supergiant stars. Others have put forward the suggestion that they are simply old novae, the irregular outbursts such as have been observed for V Sagittae and Z Andromedae bearing a great similarity to those noticed for example in the case of Nova GK Persei (1901) and many other old novae.

Here we may perhaps gain a better insight into the problems associated with the unravelling of these complicated light curves by considering one particular variable.

V Sagittae

This variable has received much attention from both photometric and spectroscopic investigators since its discovery in 1902. Despite all of the attention paid to it since the beginning of the century it has not been found possible to classify it as anything but irregular but since it complies with the conditions we have laid down for inclusion in the class of symbiotic stars, it will be discussed here and, as we shall see, all of the evidence points to it being an eclipsing binary system.

The early investigators agreed on the irregular nature of the light variations but were also able to define two apparently unrelated activities; quite rapid fluctuations of about two magnitudes with a period of a few days and a far longer variation of between three and four magnitudes taking place with a cycle of several hundred days, the latter outbursts being of the abrupt, nova-like variety.

More recent and comprehensive observations made with larger instruments have confirmed the minor fluctuations which occur with a period of approximately seventeen days and also the nova-like eruptions due to an explosive ejection of material from one component of the system. In addition to these, examination of high-dispersion spectrograms and photoelectric observations on the U, B, V system have revealed that the variable is certainly an eclipsing binary with a period of 0.514195 day. The situation is not, however, as simple as with the majority of close binary stars previously discussed in the Section on extrinsic variables for although a deep primary minimum and a much shallower secondary minimum are clearly shown, the shape of the light curve varies from one night to the next and rapid fluctuations of about 0.1 magnitude are superimposed upon the mean curve.

Here we must look at the effect of the longer variation on the eclipsing light curve. Observation shows that the amplitudes of the minima are greater when the star is passing through the minimum of this long variation than when it is bright and consequently much of the information concerning the two components of this system has been gained while the star is faint.

The picture we have at present of this remarkable system is one of two highly luminous stars, the primary being the brighter and less massive of the two. Neither star completely fills its lobe of the inner equipotential surface although the primary comes very close to doing so under 'normal' conditions. From the radial velocity curves, the mass of the primary is estimated as 0.75 times that of the sun, that of the secondary being about 2.80 solar masses. Their radii are slightly smaller than that of the sun, the secondary being somewhat the larger. The secondary component too would appear to be surrounded by a gaseous envelope very similar to that in CQ Cephei.

So far we have made no mention of the cause of the nova-like eruptions which occur at intervals of more than 100 days. If we accept the above model for this system, then it may be assumed that at certain irregular intervals the primary component suddenly brightens, throwing out a cloud of gas which, since this star almost fills its lobe of the inner equipotential surface and the material is ejected with very high velocities in excess of 400 kilometres per second, envelops the entire system.

Perhaps the most peculiar feature of this system, and the one which has an important bearing on the possible evolution of V Sagittae, is the calculated surface temperatures of the two stars. Assuming that these stars radiate essentially as black body objects, the surface temperature of the primary works out at 44,000°K and that of the secondary 22,000°K.

How can we possibly explain this high degree of over-luminosity, the primary for example being about eleven magnitudes over-luminous if we compare it with similar dwarf stars of the same mass? In the case of Algol and β Lyrae variables we saw that both components were formed at approximately the same time and for the same reasons we must accept that this is also so for the symbiotic variables. At present the only explanation which can be given for these very high luminosities is that both stars, but particularly the primary, have attained an extremely advanced stage of evolution. In other words, helium burning has reached the point where the energy-producing skin around the core is very close to the stellar surface, leaving only a thin hydrogen layer which has not yet reacted. It is quite likely that further work on these stars will produce modifications to the picture we have just outlined. At present it is extremely difficult to find any other kind of star, no matter how highly evolved, with which to compare these objects.

P Cygni Stars

These variables have sometimes been termed the permanent novae since there is evidence from their spectra that mass is constantly streaming outward from their surfaces although, of course, not with the tremendous velocities found in the novae at the time of an outburst. The most noticeable feature of the

P Cygni stars appears therefore in their spectra and not in any light variations. The latter are almost invariably slow and irregular, rather like that of η Carinae although only amounting to a few magnitudes and with long, intermediate periods when the star remains at virtually constant brightness. P Cygni itself was first recorded as a third magnitude star about 1600 but six years later began to decline slowly until it was invisible to the naked eye by 1626. In 1654 it brightened once more to just above sixth magnitude and after numerous minor fluctuations reached 5.2 magnitude in 1715 where it has remained ever since.

The most striking characteristic of the spectra of these stars is the presence of one or more emission lines with absorption components on the violet borders. The absorption spectrum underlying the emission lines is one of a typical O or B type star. The P Cygni profiles are normally seen in a few of the Balmer series of hydrogen although in some stars they also occur in the helium lines. It is now generally accepted that there is a continuous ejection of mass from these stars resulting in a large, expanding atmosphere. The absorption spectrum originates within that part of the extended shell which lies between us and the stellar surface whereas the emission lines come from gases on either side of the star, regions through which the underlying continuum does not pass. Since this large, diffuse atmosphere is expanding, there is a Doppler shift towards the violet associated with the absorption lines.

The majority of the P Cygni variables are very bright with visual absolute magnitudes as high as -7 to -9. A few are known which possess a lower luminosity than this and here it is noticeable that the absorption lines are sufficiently broadened to indicate a high rate of rotation. This observation would tend to show that such a high axial revolution may in some way contribute to the ejection of mass from the surface of the star.

24. Secular Variables

In the course of this book we have so far discussed those variable stars whose light changes can be observed over a reasonable period of time although admittedly in certain cases several years may elapse (as with some of the irregulars and the R Coronae Borealis stars) during which barely detectable variations occur. There are, however, some of the brighter naked-eye stars in which changes in brightness appear to have taken place over several hundred years.

At the present time, only one or two of these secular variables, as they are termed, show any fluctuation in light, variations which seem to be completely irregular and of extremely small amplitude. Whether we are justified in assuming that any change at all has really occurred in the magnitudes of these stars is a debatable point since in almost every case we are forced to accept the accuracy of the magnitudes as given by Ptolemy in the Almagest produced in the second century A.D. and based upon an earlier catalogue by Hipparchus. It is quite possible that errors of observation were made by these early observers but since these amount to more than a magnitude, the stars concerned will be mentioned here for the sake of completeness.

One of the best known examples is the star Pleione in the Pleiades which is now appreciably fainter than its close companion Atlas. In the Almagest, both are given as being of the same magnitude. It would appear therefore that this star has faded by more than a magnitude in the past two millennia. Other stars which have apparently faded by similar amounts include δ Ursae Majoris, α Geminorum, β Librae, θ Eridani and β Leonis. Of these, the last two were recorded as first magnitude stars by Ptolemy.

There are, too, examples of stars which have brightened over the intervening period such as β Canis Majoris (a star which has been discussed elsewhere in this book and which is known to be variable) and α Ophiuchi.

Although we must be very cautious in our acceptance of the

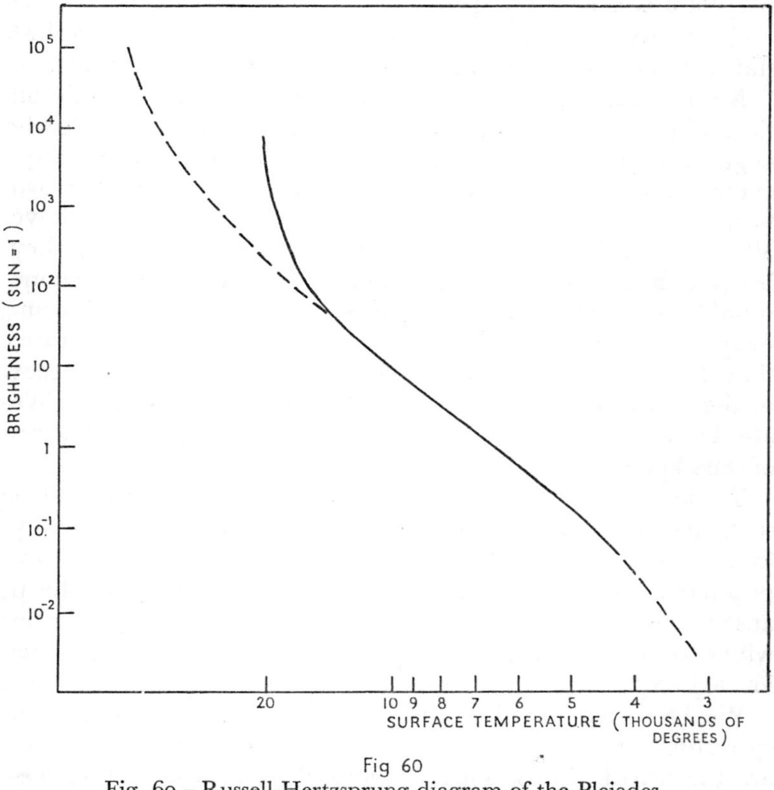

Fig. 60 – Russell-Hertzsprung diagram of the Pleiades.

reality of these discrepancies, it seems unlikely that we can dismiss them all as being due merely to errors of observation. Such variations, although on a shorter time scale, have been found for the P Cygni stars and here it is of interest to note that Pleione is a peculiar star showing certain similarities to the P Cygni variables. It is perhaps worthwhile considering this star in relation to the other stars of the Pleiades.

The Pleiades are perhaps the best known of the open clusters, the members being bright, young stars, all of which may be assumed to be of about the same age, unlike the globular clusters which are of much greater antiquity and far more densely populated. Being young stars we might expect that if we plot a Russell-Hertzsprung diagram for the members of the Pleiades, we shall find some indication of stars which are just starting to evolve. This is exactly what we do find (Fig. 60).

Most of the stars in the Pleiades cluster are found to lie on the main sequence but others, in particular Pleione, have clearly begun their evolution to the right. Does this then, in any way, explain the fading of this star by something approaching two magnitudes? Quite clearly, the answer is no; since we have already seen that as main sequence stars begin to evolve, they become both larger and brighter and consequently Pleione should have increased in brightness. Now we know, both from its spectrum and position on the Russell-Hertzsprung diagram, that this star has a diameter about five times greater than that of the sun and a surface temperature close to 15,000°K. We also know from a study of its spectrum that it belongs to a class of stars known as shell stars.

These stars are found exclusively among the types B and A stars and show broadened lines of all the elements normally found in stars of these spectral classes. The lines of hydrogen are peculiar in that they consist of broad lines having very sharp absorption cores and in addition there may also be emission wings to these lines. Generally, only the absorption lines of the ionised metals tend to be sharp and narrow.

We can therefore distinguish three components of such a spectrum depending upon the origin of the lines. Firstly there are the broad, diffuse lines of hydrogen and a few other elements which arise within the atmosphere of the star itself, a star which is obviously rotating very rapidly. Surrounding such a star and at some distance from it, there is a large shell of gas which, because it is sufficiently removed from the surface of the star, rotates with a much lower velocity. Where emission lines are present in the spectrum, these originate in this shell of gas. Similarly, the narrow absorption cores of hydrogen and the lines of the ionised metals arise in that part of the shell which lies between us and the photosphere of the star acting as a kind of reversing layer. That the shell actually is rotating more slowly than the star is shown by the fact that those lines which originate here are not broadened to any extent by rotation. Small variations in brightness are found in several of these shell stars but not to the extent of two magnitudes. The fading of Pleione is therefore still a mystery and no satisfactory explanation has yet been given.

Such is also the case with the other secular variables. Here there is even less indication of any peculiarity associated with these stars. In the case of δ Ursae Majoris, small and irregular fluctuations have been suspected for many years and it may possibly be an irregular variable subject to long periods of constant brightness.

25. The Role of Variable Stars in Astronomical Research

Up to this point we have confined ourselves mainly with a discussion of the properties of the various kinds of variable stars, how certain classes are interrelated, their probable modes of evolution and their spatial distribution. We now come to a feature of these stars which we have mentioned only briefly. The central idea of this chapter can be put very concisely. Taken as a whole, variable stars cover the entire range of spectral types, surface temperatures, sizes and densities; they include members of both Population I and Population II stars, those which are still in the process of condensation from the parent gas cloud and others almost as old as the galaxy itself. They have been identified in large numbers in those galaxies which form the Local Cluster and some (the novae and supernovae) in even more distant nebulae. As a result, they have proved of immense value in many fields of astronomical research providing us with a much better understanding of the physical and nuclear reactions going on inside the stars and, going even further afield, vital information about the dimensions and structure of the galaxy and the universe of the extragalactic nebulae.

In the present chapter therefore the aim is to consider in more detail the ways in which a study of variable stars has helped us to formulate much of our present-day knowledge of the subject.

During the first two centuries which followed the discovery of the variable nature of o Ceti and β Persei, astronomers concentrated on a broad classification of these stars as more and more were identified. Light curves were produced for a rapidly growing number of variables but for only one type of star, the eclipsing binaries, was a correct explanation of the light variability forthcoming. The need for a systematic nomenclature was apparent quite early in the study of variable stars and the simple system devised by Argelander and extended in 1881 by Hartwig, was finally adopted.

Towards the end of the last century, the emphasis gradually shifted away from the production of visual light curves as more powerful instruments and refined techniques came into use and the introduction of spectroscopic observations was a major milestone in our understanding of these stars. Such information provided the first proof of Goodricke's suggestion that the Algol variables were binary systems, the light changes being due to the mutual eclipse of the two components during each revolution around the common centre of gravity. In addition, they lent strong support to the pulsation theory of the Cepheids and the long period variables as well as demonstrating the ejection of gas at tremendous velocities from the novae and supernovae.

The Basic Properties of Individual Stars

Certain features of the stars may, of course, be deduced simply from their spectra whether they are variable or not. We can form a fairly accurate estimate of their surface temperatures for example and from the relative intensities of the absorption lines tell whether a star is a giant or a dwarf. The width of the lines also provides a means of estimating the velocity of axial rotation.

It is when we come to measure such characteristics as the diameters, masses and densities of the stars that we run into difficulties. The interferometer can be employed for the measurement of stellar diameters only for nearby supergiant stars such as Antares and Betelgeuse; the vast majority of stars are either too small or too distant for this method to be used with any hope of success. Fortunately, in the eclipsing variables another direct method is available, one which is not so dependent upon size or distance.

The eclipsing binaries provide us with an ideal, direct method of determining the masses of the components since their motion around the common centre of gravity conform to the law of universal gravitation. Such masses can, of course, be readily deduced from the orbital elements of any double or multiple system (although where stars are concerned it is somewhat more difficult in the latter case) and also, visual doubles must be observed for relatively long periods, first to detect any motion which may be present and secondly to derive the elements of the orbit. With the Algol and β Lyrae variables we do not have this difficulty. Their periods are normally quite short, generally of the order of a few hours or days. Certainly the number of such binaries for which accurate orbital elements are known is somewhat limited (between one and two hundred in all) but since we find that virtually all stellar types are represented, we are able to calculate the masses of a wide variety of stars by this method.

One further great advantage of these variables is that the light changes which accompany this binary motion provide us with additional information which cannot be directly derived from the visual or purely spectroscopic binaries. The duration of the eclipses, taken in conjunction with the velocity of each component in its orbit (found from the radial velocity curves), enables us to find directly their diameters and hence their densities.

One somewhat unexpected result which emerged from a study of these systems and which has been extended to cover all types of stars is that although there is a tremendous variation in luminosities and diameters, the masses of the stars fall within very narrow limits. At the lower end of the scale there are dwarfs with masses about one-tenth that of the sun while the upper end is populated by supergiants having masses between ten and fifteen times greater than the sun. Only in very rare instances do we find stars having masses outside these limits.

Since there is such a wide variation in stellar diameters, then we must expect an equally wide spread of densities. The white dwarfs which are found among the old novae for example have densities more than 100,000 times that of the sun, the matter of which they are composed being compressed to an almost unbelievable degree so much so that the electron shells themselves must have collapsed under the tremendous pressures existing within these stars. The supergiants on the other hand have densities so low that they may be regarded to all intents and purposes as approximating a good vacuum here on Earth.

Proof of the extremely tenuous nature of the atmospheres of these stars is provided by such eclipsing binaries as ζ Aurigae where we can actually observe the light of the bright B8 component shining through the extended atmosphere of the K4 supergiant star at the beginning and end of the primary eclipse. Admittedly there are only a few such eclipsing variables where we are able to directly observe the atmosphere of a star but such data has greatly extended our knowledge of the composition and density gradients of stellar atmospheres.

There is a further important contribution which these close eclipsing binaries have made to astrophysical research, namely that of mass loss from a star during the course of its evolution. The sun is losing mass continually in the form of radiation, the energy coming largely from the proton chain reaction occurring deep within the sun and from the carbon-nitrogen cycle, both of which result in the conversion of hydrogen into helium as we have already seen.

When we come to consider close binaries, the position is found to be far more complex. Nevertheless, we may make some generalisations based upon observation. Firstly, mass may be lost very slowly in the form of radiation as in the case of the sun, a suggestion originally put forward by Jeans some forty-five years ago. Provided that this radiation is emitted spherically and with a sufficiently high velocity there will be no change in the elements of the orbit of a binary system. It is very unlikely however that this ideal situation occurs. As Blaauw has shown, if ejection under these conditions takes places, not continuously, but instantaneously, then there will be complete dissociation of the two components if they are moving in a circular orbit. If they are revolving in an elliptical orbit, then according to Huang the two stars will coalesce into a single object as long as the mass ejected exceeds a critical amount which depends upon their masses and separation at the time of ejection.

What will happen however if the mass is ejected with slow velocities? Obviously it will not, under these circumstances, escape from the system. Two things may then occur. Either the mass will fall back into the parent component or it will be captured by the companion – or it may form a ring of gas around the more massive of the two stars. Such rotating rings of gas have been observed in several binary systems and have been discussed in earlier chapters, providing a striking confirmation of this mode of mass ejection.

Both of the above modes represent somewhat idealised extremes. We must now consider the case where the mass is ejected with velocities intermediate between the two. Here the ejection velocity is assumed to be sufficiently high for mass to escape through the inner equipotential surface of the system but not from the system as a whole. Any ring which may then be formed will encompass both components and to be dynamically stable will have to be at a large distance from either star. Again, there is observational evidence from some binary systems in support of this.

As we might expect, it is extremely unlikely that there are any binary systems in which one of the above modes of mass loss takes place to the total exclusion of the others. Atoms and molecules will be emitted with a wide range of velocities and there will also be differences in the angular momentum per unit mass.

Now although it is possible for us to infer the presence of moving streams of gas and rotating rings by means of the spectroscope and from irregularities in the photoelectric light curve (particularly if the system is observed by three-colour photometry), are there any other effects which such mass losses will produce in the system? One of the parameters where we might expect to find some change is the period of orbital revolution. Several eclipsing binaries are indeed known in which the period is slowly changing and as Huang has shown, in the case where there is mass transfer from one component to the other without the formation of a ring around either star, the period will increase where transfer takes place from the less massive star to the more massive companion and will decrease where the mass transfer is in the opposite direction. It is gratifying to find that examples of binary systems are known corresponding to both cases.

Suppose now that a rotating ring of gas is formed around one of the stars. Such a gaseous ring around the less massive component can be shown to be dynamically very unlikely. Almost certainly it does not happen. In those circumstances where the ring is formed around the more massive star, it can be demonstrated theoretically that the period may either increase or decrease but in any event, such changes will be much smaller than those produced simply by mass transfer.

Now what of the case where a large ring is produced around both stars? Here we can show that such a ring formation will lead to a decrease in the period. The results we have thus obtained prove to be extremely useful in determining the principal mode whereby mass losses occur. An excellent example of this is β Lyrae, the period of which is slowly and smoothly increasing. From this we may conclude that there is a transfer of mass from the less massive component to the more massive star, a conclusion which has had a profound effect on our ideas of the evolution of this system. In Chapter 2 we saw how, if this is the case, the primary which is now the less massive of the two must originally have been the more massive. The increase in the orbital period is direct and independent confirmation of this view. Since there is also a shell of gas surrounding both stars in this particular system, mass is clearly being lost by two modes, but since there is a slow increase in the period, mass transfer from one star to the other is clearly the main method by which mass is exchanged.

One other feature of these close binaries has yet to be mentioned here. As we saw in the first three chapters, synchronisation between the orbital and axial revolutions occurs in certain systems and is absent in others. How does such synchronisation come about? Again we find that mass transfer, if not wholly responsible, provides a major contribution towards an understanding of this problem. If we attempt to explain this in quantitative terms we find the problem to be extremely complex. Nevertheless, considering the effects of tidal forces between the stars and mass ejection we are able to obtain a somewhat simplified picture which provides us with a general answer to the problem.

In the case of the W Ursae Majoris variables where the component stars may, to all intents and purposes, be regarded as being in physical contact, synchronisation must occur owing to the magnitude of the tidal forces. However, in all other cases, where the individual stars may be regarded as discrete objects, we find that the time scale over which mass ejection occurs is a dominant factor. If this is sufficiently long to enable tidal and gravitational forces to exert a considerable effect, thereby coupling orbital and axial rotation, then the stars will be synchronised; on the other hand, if mass transfer occurs over a very short period, then we would expect orbital and axial motion to be non-synchronised.

The Birth of Stars

The problem of stellar evolution has occupied astronomers for more than a century now and it is only recently that a reasonably satisfactory picture has been built up which explains most of the observed facts. Yet towards the end of the nineteenth century when Lockyer put forward his evolutionary scheme it appeared that little more remained to be said on the subject. Briefly, Lockyer postulated that a star began its career as a huge gaseous mass condensing from a gas cloud. As the star contracted, the temperature rose and in successive stages the star became a red supergiant, an orange giant, a yellow giant and finally a blue-white star as the temperature reached its maximum. This was then the turning point in its evolutionary career, for now the star began to cool as the rate of contraction diminished, passing down the main sequence through the yellow dwarf stage (as the sun is now) and ending its career as a red dwarf before final extinction.

Such a sequence was not only straightforward but appeared to receive additional support when some years later the division of the stars into giants and dwarfs (which had not been recognised at the time Lockyer put forward his theory) was made, mainly on spectroscopic evidence. When the lifetime of such a star, based upon purely gravitational energy, was calculated, however, it was found to be only of the order of fifty million years, a time scale far too short to explain the age of the sun which was shown, from several independent sources, to be in excess of three thousand million years. Lockyer's scheme, although extremely plausible, was clearly untenable.

Some years later, Russell suggested that although the sequence of events put forward by Lockyer were correct, the source of energy was not gravitational contraction, but due to nuclear processes occurring inside the star. This removed the objection of the too-short time scale only to replace it by another, equally disastrous. Now the time scale was far too long! Russell's suggestion placed the age of the sun at not less than nine million million years.

We now know that it is the sequence itself which is wrong. A star passes through the yellow dwarf stage, like the sun, comparatively early in its career. After condensing out of the parent gas cloud, the star evolves towards the main sequence joining it at the lower end at a point which depends upon its mass and possibly on whether the material of which it consists is mixed in any way. After passing up the main sequence, growing brighter and hotter as it does so, it then evolves to the right provided that little or no mixing occurs, becoming larger and redder in the process eventually evolving back towards the main sequence and finally ending up as a white dwarf with or without the intervention of the nova phase.

One of the most important and elusive phases of a star's evolution is that before it evolves onto the main sequence just after the initial condensation process. In the very first stages, the star will emit no visible radiation since the temperature is still not sufficiently high for it to do so and it is perhaps significant that in recent years several dark globules have been discovered projected against the background of bright gaseous nebulae which may be protostars just at the point where they are beginning to condense.

The importance of certain kinds of variable stars is that there is now a growing amount of evidence to show that they are extremely young stars passing through the phase between condensation and reaching the main sequence. The logical birthplace of the stars is, of course, deep inside the gaseous nebulae such as that in Orion and in the vast associations which are found in Auriga, Taurus, Orion and Monoceros. In many cases, as we have seen earlier, the surrounding gas and dust obscures these objects completely and it is for this reason that a study of the RW Aurigae, T Orionis and T Tauri stars has proved invaluable.

Many of these variables emit powerful infra-red radiation as we would expect from stars just contracting from the pre-stellar gas and dust clouds and their spectra, consisting of both emission and absorption lines, together with their irregular light variations make them relatively easy to identify.

Undoubtedly future investigation of these variables will provide the answers to several problems concerning the initial stages of a star's career. Certainly the time scale for contraction onto the main sequence is very short compared with that for the succeeding evolutionary stages and in support of this there is the observation that at least three irregular variables of the T Orionis type have been photographed in the region of the Orion Nebula which apparently did not exist in stellar form at least a decade earlier. In addition, an infra-red source has been discovered within the nebula which, by analogy, might conceivably appear as a T Tauri variable within the next two decades. We must, of course, be wary of reaching any definite conclusions concerning these objects in the present state of our knowledge of these variables.

Variable Stars as Distance Indicators

One of the great advances in astronomy has been the accurate measurement of interstellar and intergalactic distances. Until the latter half of the nineteenth century all attempts to measure the distances to the stars ended in failure and it was only possible to say that since no motion among the stars could be detected they lay at enormous distances from the solar system. The first method to be used – that of trigonometrical parallax – consists of measuring the parallax of a star at six-monthly intervals, since this provides us with the longest base line possible, namely the diameter of the Earth's orbit around the sun, approximately 186,000,000 miles. It was by this means that Bessel, in 1838, measured the distance to 61 Cygni. This method, of course, suffers from one very serious disadvantage. It can only be used for stars which are very close to the sun. Beyond a distance of about 300 light years it fails utterly even when the largest telescopes available are used to measure these extremely small angles. Were it not for other methods it would be quite impossible to make accurate measurements beyond this distance.

Clearly we require some standard stellar light source by

means of which we can estimate distance directly since from the inverse square law we know that the apparent brightness of a star is inversely proportional to its distance. The properties required of such a star are that it should be intrinsically as bright as possible (to enable us to see it at great distances) and it should possess some distinguishing feature which will make it possible for us to pick it out without difficulty from the myriad of other stars. Fortunately the RR Lyrae variables satisfy both of these conditions. Their light curves are quite characteristic and readily recognised and in addition it has been shown, as described in Chapter 5, that they all have the same absolute brightness, being about a hundred times more luminous than the sun.

From a statistical survey of these variables in the galactic disc, Baade has shown that the centre of the galaxy lies at a distance of almost 30,000 light years from the sun. Away from this point, the frequency with which the RR Lyrae stars occur along the Milky Way falls off quite appreciably. One further advantage of using these variables as distance indicators is that they are found in large numbers within the globular clusters which form a sort of halo around the galaxy. We can therefore use them to determine the distances of these clusters and from these results obtain both the shape and dimensions of the galaxy. We have seen that some discrepancies arose owing to the erroneous belief that the absolute magnitudes of these stars are about 0.0, a figure which was later modified to +0.8. More serious trouble was encountered when the Cepheids, which are intrinsically even brighter than the RR Lyrae variables, were used for the determination of distances outside the galaxy. Not until these two types of variable were recognised as belonging to two different classes of stars was this difficulty finally resolved.

Even the Cepheids are not sufficiently bright when we come to consider the more distant extragalactic nebulae and more luminous stars (the long period variables and the novae) have been employed for distances out to about thirty-five million light years. Beyond this tremendous distance, individual stars cannot be identified, either visually or spectroscopically, with the rare exception of the supernovae. As a final resort, we must use the nebulae themselves as standard light sources although such results will naturally be less precise than those based upon individual stars.

To end this chapter we must say something about the supernovae themselves since they have provided us with a very plausible answer to one of the fundamental problems facing astrophysicists, namely the origin of the heavier elements in the stars. If we look at this point a little more closely we shall see just how important this question is. If we assume that a star originally condenses from a vast cloud of hydrogen then as we have seen, the predominant nuclear reactions which occur are those of the proton chain process by means of which the hydrogen is built up into helium. This is the major reaction sequence taking place in the main sequence stars, the carbon-nitrogen sequence being secondary at this stage. Spectroscopic examination shows, however, that these stars also contain a small proportion, about one per cent, of heavier elements, in particular those from carbon to zinc in the periodic table. Now where did these heavier elements come from? Since there are no nuclear reactions going on in these main sequence stars to produce them, clearly they must have arisen by some other process.

The inescapable conclusion is that if they have not been produced inside the star itself, then they must have been present in the gas from which the star originally condensed. For many years this presented astronomers with a seemingly insurmountable problem. It appears very unlikely that the primordial gas contained even this small amount of these elements. Consequently some mechanism must have been available for its production quite early in the history of the galaxy. The answer which is accepted by many astronomers is that these elements were initially produced in the interiors of supergiant stars having masses larger than that of the sun. This only partly answers the problem however, for so long as these elements remain inside the star they cannot become mixed with the gases of the galaxy from which we might expect later generations of stars to be born. How then were these elements released from these stars? The answer appears to be in the process of a supernova explosion.

As we have seen in Chapter 22, the Type II supernovae, being Population I stars, are among the oldest of the stars and also have masses much greater than that of the sun. It is also in such stars as these that the heavier elements are formed during the nuclear reactions which lead to the eventual explosion. We have also seen that there is observational evidence for the existence of Type I super-supernovae occurring during the very early stages of galactic evolution, stars with incomparably large masses. As the gases from such explosions, ejected with very high velocities, became dissipated throughout space, it would mingle with the primeval hydrogen so that stars which then condensed at a later stage would contain a certain proportion of these elements in their original composition.

As we may imagine, it is not an easy matter to make any quantitative comparison between the total amount of these heavier elements which has been produced by supernovae explosion during the lifetime of the galaxy and that which is present in the main sequence stars at the present time, but the results which have so far been obtained show a fair agreement which is perhaps the best we can hope for in the circumstances.

The result of this discussion is that variable stars as a whole provide the astronomer with the means of attacking many astrophysical problems. In some cases the interpretation of the observations is difficult and liable to error but it appears that further research on these stars should be able to answer many questions still posed by astronomy.

26. The Observation of Variable Stars

One of the main aims of this book has been to present an overall picture of our present knowledge of the many different types of variable stars now recognised by astronomers. In many cases, of course, it is difficult to place a star into any particular class and the dividing lines between some of the groups are by no means sharp, indicating perhaps certain fields of variable star research where further work must be done before any definite conclusions can be reached.

It appears unlikely, though by no means impossible, that further types, different in any appreciable degree from those variables already described, remain to be discovered. It is probably true that almost all stars are variable to a certain limited extent; we may argue that our own sun is a variable star owing to the well-known sunspot cycle of about eleven years but the resulting change in luminosity is so infinitesimally small that it cannot be classed as a variable star in the accepted meaning of the term.

No book on variable stars would be complete, however, without some reference to the various methods used by observers in their investigation of these very important and interesting stars. For more than two centuries all of the data obtained was purely visual and the estimates made were inevitably far less accurate than those made today using large instruments and refined equipment and techniques. Nonetheless, they must not be despised because of their limitations.

Our present-day knowledge of variable stars rests on the foundation built up by these early astronomers, many of whom were amateurs. The major difficulty they faced was that of obtaining accurate and consistent sequences of comparison stars against which to estimate the brightness of the variable and it is only comparatively recently, in 1922, that the universally adopted systems of visual and photographic magnitudes based upon the North Polar Sequence has led to the provision of a standard against which all other magnitude determinations may be made.

In the first part of this chapter we shall discuss the role which the amateur with the limited equipment at his disposal may play in the observation of variable stars and then go on to mention the modern methods employed by the professional astronomers. At first sight, it might appear that visual estimates are necessarily inaccurate and of little use: nothing could be further from the truth. Provided one recognises the obvious limitations of this method, extremely useful results can be obtained. Naturally there are certain classes of variable stars which are quite unsuited to the purely visual technique. The extrinsic variables, except for those of fairly long period such as ζ Aurigae, can only be satisfactorily observed photoelectrically since their periods are so short and their amplitudes so small that even small errors in estimation of brightness tend to give quite erroneous results.

When we come to the intrinsic variables however the position is quite different. Only the RR Lyrae stars, short-period Cepheids and β Canis Majoris variables are better observed throughout their light cycles by the more sophisticated techniques. The continuous observation of the long period variables, the semi-regulars, U Geminorum variables, RV Tauri and R Coronae Borealis stars has long been in the hands of the dedicated amateur. Of necessity the available time of the large telescopes in the main observatories of the world must be utilised to its best advantage, namely the investigation of the many cosmological problems facing astronomers, particularly those associated with extremely faint and remote objects lying outside of our own galaxy.

Finding the Variable

One of the major difficulties experienced by beginners to variable star observing is that of finding the variable. Since it may be invisible at the time, the problem resolves itself into one of locating the star field rather than the star itself. The charts used are normally based on Hagen's *Atlas Stellarum Variabilium* and several charts are prepared for each star covering, in general, the 9°, 3°, 1° and 20′ fields. The position of the variable with respect to several naked eye stars will be shown on the 9° chart and by means of certain key comparison stars (those which are not variable and whose magnitudes have been accurately determined for use in estimating the brightness of the variable) will lead the observer from one chart to the next, narrowing the field at each step and introducing fainter and fainter comparison stars deliberately chosen so as to be conveniently close to the variable.

It is a definite advantage to know the field of view of the telescope being used and its magnification beforehand. An accurately set equatorial mounting with divided circles, possibly fitted with a clock drive, will find the field of the variable if the necessary calculations are made, but this is tedious and time-consuming and except in the case of very large apertures and high magnifications is not really necessary. A good knowledge of the naked eye stars and a low-power finder are all that is required for a preliminary search. If the variable is not visible at first, a higher magnification may sometimes enable the observer to pick it out since this tends to darken the background and makes the stars appear relatively brighter. Averted vision is also helpful but more will be said about this later. Once the field has been located and the variable observed on several occasions, the time taken for future identification will be reduced to only a few minutes enabling the observer to make many visual estimates in quite a short time.

If an equatorial mounting is used it is often possible to set the telescope on a known star having either the same Right Ascension or Declination and identify the field of the variable by means of a short sweep in the appropriate direction. Where the observer is equipped only with an altazimuth mount, the method of 'drift' is then to be preferred. A known star is chosen which precedes the field and has approximately the same Declination as the variable. Using a low power the telescope is then aligned on the known star and the time of observation noted. When the time interval between it and the variable has elapsed, the latter should be somewhere close to the centre of the field of view.

A word might be said here about the performance of various telescopes for variable star observing. As one might expect, the larger the aperture, the more variable stars may be observed and particularly in the case of the long period variables and the U Geminorum stars, such large apertures are necessary to follow these stars through their faint minima. When at their maximum however a large number of variables are between second and seventh magnitude and are best observed with binoculars at this phase. There are two reasons for this; firstly one must avoid excessive brightness since this leads to an uncertainty in the estimation of magnitude especially of red stars and secondly it will be appreciated that comparison stars at the bright end of the sequence will be situated further from the variable than those at the faint end and a more accurate estimate can be made when one does not have to swing the telescope any great distance from one star to the other. We must avoid the temptation however to try to view both variable and comparison star simultaneously, since, as we shall see, this can introduce errors in estimation.

Large apertures are naturally most useful on very faint stars and should be largely devoted to this work. Stopping down the aperture may be done with advantage as the variable brightens and this will also serve to flatten the field with low powers. Experience has shown that there is an optimum optical power for any given magnitude of the variable, this being the one which will show stars about three magnitudes fainter than that which is being observed. Any wider separation in brightness than this is not generally desirable.

Estimation of Brightness

The most important and numerous group of stars observed by amateurs at the present time are the long period variables. Not only do they possess large amplitudes but their periods are measured in hundreds of days and consequently the inevitable 'scatter' of estimates about the mean is less serious in the production of the light curve than it would be in the case of the other types of variables. Unfortunately, the long period variables do have one particular drawback as far as visual observation is concerned, namely that they are all red stars, and later we shall see that this introduces a further difficulty in the estimate of their brightness. Generally speaking, it may be said that with experience, two independent observers will agree to within 0.1 or 0.2 magnitude in their estimates on any particular night for white or yellow-white variables, e.g. the U Geminorum or RV Tauri stars; but for red stars the difference in estimates may be as much as one or two magnitudes.

To understand the reason for this wide discrepancy we must here discuss the question of colour acuity and in particular the so-called Purkinje Effect which accounts largely for the wide variation of results obtained by observers of the strongly coloured Mira type stars.

When two point sources of light, one red and one white, of equal intensity are brightened by the same amount it will be found that the red light is now the brighter of the two, the converse being true when they are dimmed equally. This effect is due to the fact that the fovea of the eye, when light adapted, becomes relatively more sensitive to red light than to white. It is, of course, possible by various means to minimise the Purkinje Effect but not to eliminate it altogether.

One method which immediately suggests itself is the use of colour filters. A disadvantage of these is that they inevitably reduce the amount of light reaching the eye and are therefore of little use in the case of faint stars. To obtain uniformity of visual estimates such filters would have to be used consistently throughout the whole of the light cycle and would seriously restrict the use of a given telescope around minimum where the light absorption due to the filter could amount to a magnitude or more.

Another way is not to look directly at the star but to allow the light to fall on the outer part of the retina. Again, we come up against a difficulty since it is well known that the outer parts of the retina are relatively insensitive to red light, making a red star appear fainter than it should, and perhaps the best method of all to use when red stars are particularly bright is to stop down the instrument or use a smaller telescope. It is often found by observers when a red variable is brightening towards maximum that on changing from a telescope to binoculars there is a distinct discontinuity in the estimated brightness almost certainly due to the abrupt reduction in the aperture being used.

Coming now to the actual estimate of brightness, two methods are in general use by visual observers. In both, the variable is compared with, if possible, at least two of the comparison stars so chosen that one is slightly brighter and one slightly fainter than the variable. These are:

(a) The Fractional Method: Here the brightness of the variable is mentally placed between those of the two comparison stars and expressed as a fraction of their difference in magnitude. As far as possible, the sequences of comparison stars are chosen such that the difference in brightness between any two does not exceed half a magnitude. An example will illustrate this method more clearly.

We shall suppose that an observer has mentally divided the brightness interval between two comparison stars A and B into five equal parts and estimated the variable as one fraction fainter than A and therefore by implication four fractions brighter than B. The notation for such an estimate is A(1)V(4)B. It will be noted in this and the following paragraphs that the brighter of the two comparison stars is always given first.

If the magnitudes of the comparison stars are, say, 8.76 and 9.32 respectively, then the brightness interval is 0.56 magnitude and one fifth of this is 0.11 magnitude. Adding this to 8.76 gives us 8.87 for the deduced magnitude of the variable. It will be seen that the same result is obtained by subtracting 0.44 from 9.32, the small discrepancy of 0.01 magnitude being insignificant in this case.

It may be that occasionally the variable cannot be placed between any two of the comparison stars either because it is abnormally faint or extremely bright and in this instance two stars of not very different magnitude may be found, both either brighter or fainter than the variable. We may thus have an estimate of, say, A(2)B(3)V, implying that the variable is half as faint again as B, as B is fainter than A. If we assume the same magnitudes for the comparison stars as before, then the variable is 0.56 x 3/2 or 0.84 magnitude fainter than B. Simple arithmetic then gives us a magnitude of 10.16 for the variable. If, on the other hand, the estimated result is V(2)A(3)B, meaning that the variable is brighter than A by 2/3 of the amount by which A is brighter than B, then it will be 0.56 x 2/3 or 0.37 magnitude brighter than A and subtracting this from 8.76 we get 8.39 as the magnitude of the variable.

In the final case where the variable is considered equal to one of the comparison stars it should always be compared to at least one other star since estimates based on comparison with only one standard are likely to be inaccurate. In this, and the following method, it is not necessary to know the magnitudes of the comparison stars at the time of making the estimate; indeed it is better not to know their magnitudes since this may, in certain cases, lead to personal bias.

(b) Pogson's Step Method: This method is not recommended to the beginner since it requires that the eye be trained to estimate differences in steps of 0.1 magnitude. It has several advantages over the previous method; the variable is being compared with only one comparison star at a time and it may also enable an experienced observer to detect variability in a comparison star which has not previously been suspected. The method should only be used after experience in variable star observing has been gained over several months and absolute reliance should not be placed upon such an estimate unless more than one comparison star has been used. Quite often an observer may use both methods in making an estimate and when this is done it is usual to give double weight to the deduced magnitude obtained from the Fractional method.

Again, as an example we may take two comparison stars A and B having adopted magnitudes of, say, 8.30 and 8.80 and assume that the observer estimates the variable as two steps fainter than A and two steps brighter than B. This is written as A−2, B+2, giving us 8.50 and 8.60 as the individual values obtained from these two estimates. The mean of 8.55 is the deduced brightness of the variable.

Lastly, we must consider the situation where the variable is not visible in the instrument being used. Here the observer records the magnitude of the faintest comparison star which can be seen, the variable being recorded as 'invisible', less than the magnitude of this particular comparison star. Such negative observations can be extremely valuable, especially in the case of the U Geminorum and Z Camelopardalis stars since they provide evidence that a maximum has not been missed owing to a paucity of observations at that particular time.

Estimation of Colour

Visual estimates of the colours of variable stars cannot be considered as exact but are useful in following the changes of novae for example and for the long period variables of classes R and N which become progressively redder as they approach minimum. Estimates of colour are often made on the following scale although it must be emphasised that the sky must be perfectly dark and moonless for them to be of any real value.

0 – Pure white	6 – Orange
1 – Yellowish white	7 – Deep orange
2 – Pale yellow	8 – Reddish orange
3 – Pure yellow	9 – Red with a slight orange tinge
4 – Deep yellow	10 – Pure red
5 – Pale orange	

Pure red is never seen and it is very doubtful if stars of Type 9 are known although this colour may sometimes be seen if the telescope is not perfectly achromatic.

Sources of Error

At this point it would be as well to take a look at some of the sources of error often encountered in variable star observing. The most common is mis-identification of the variable or of a comparison star. When one remembers that there may be more than a score of stars of various magnitudes in the field of view, it is easy to see how serious a difficulty this can be for the beginner. One must also bear in mind the fact that frequent references to the charts cannot be made while at the telescope since any form of light, even a faint red one such as is most often employed, tends to destroy the sensitivity of the dark-adapted eye. It is therefore necessary to be able to retain a mental picture of the star patterns in the various variable star fields. Fortunately the eye is unusually adaptive to the recognition of such patterns once the scale of the field of view has been ascertained and this does not normally present an insuperable problem.

Too prolonged viewing of a star, particularly a red one, will lead to discordant results owing to the Purkinje Effect previously discussed. Very often the first indication of this is that the star appeared to brighten appreciably in a series of discrete jumps.

Errors can often arise owing to the position angle between the variable and the comparison star. As has been shown by Waterfield and others, when two equally bright stars are being examined in the same field of view, not only does the star below the other in the field normally appear the brighter, but also the star towards the 'nose' side of the field. To eliminate this effect, both stars should be brought into the centre of the field as quickly as possible.

Bias is another possible source of error since one can rarely completely forget previous estimates or the magnitudes of the comparison stars. One must never assume that a variable will be fading, brightening or remaining stationary in spite of what the previous estimates tend to show. Large and unexpected irregularities often occur in the light curves of the long period variables and are even to be expected in the case of the irregulars.

The use of averted vision can also lead to marked discrepancies in estimates although provided that both variable and comparison star are observed by averted vision this is not a source of error in itself. When stars are very faint, averted vision must of necessity be used to see the faintest possible stars. What must be avoided at all costs is comparing the variable with a comparison star lying at the very edge of the field, a star which in addition to being observed by averted vision is most likely out of focus as well.

Finally we must say a word about atmospheric absorption. This is generally taken to be zero at the zenith but increases appreciably with decreasing latitude and becomes quite a serious consideration when observing stars within a few degrees of the horizon. Where it is necessary to observe a variable at these low latitudes, comparison stars should be chosen which are at essentially the same altitude as the variable in order to minimise this effect as far as possible. Where this is not feasible, tables are available giving the magnitude of the atmospheric absorption for different latitudes and corrections may be applied.

Calculation of Dates of Maxima and Minima

Even a cursory glance at the light curves of many types of

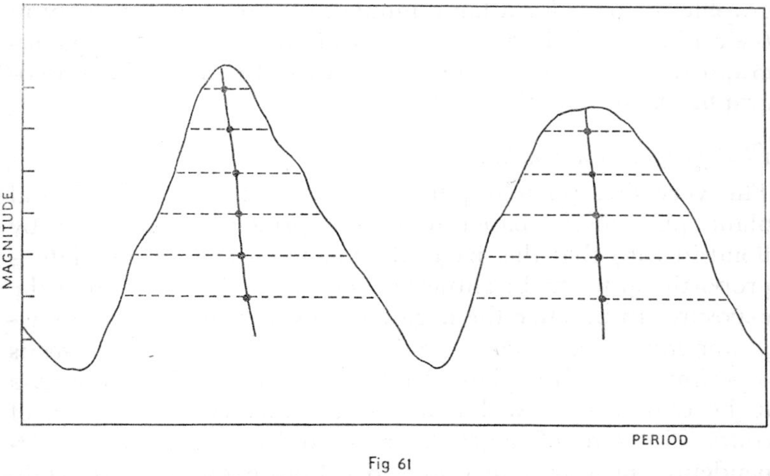

Fig. 61 – Pogson's method of bisected chords for the calculation of dates of maxima and minima.

variable star will be sufficient to show that the ascending and descending branches are not symmetrical and since there is seldom, if ever, one particular day on which a maximum brightness can be said to be attained, it is necessary to devise some means of calculating the date of maximum (or minimum), taking into account this asymmetrical shape of the light curve. The method used, again due mainly to Pogson, is that of bisected chords. From the light curve, the dates are found when the star attained the same magnitude on the ascending branch as on the descending portion and the line joining these two points is then bisected. This process is then repeated for various magnitudes and the points bisecting the chords are then joined and extrapolated to the peak of the curve. (Fig. 61).

This method is of particular value in the determination of the dates of maxima of the U Geminorum and Z Camelopardalis variables which commonly exhibit fairly long, reasonably flat maxima.

We now come to a discussion of the various methods used by the professional astronomers in their study of variable stars. Although visual photometry is perfectly adequate for many kinds of variables, it has been increasingly superseded by photographic methods for a large number of stars and especially in their discovery where wide areas of the sky may be photographed on the same plate and compared with similar photographs taken some time later.

Photographic Photometry

The very first photographic method used was that of focal plane photometry based upon the principle discovered by Bond in 1857 that the size of the image formed on the plate is proportional to the brightness of the star and the length of the exposure. In practice the negative is examined through a low-power microscope and a graduated scale of stellar images is superimposed on the plate enabling the magnitudes of the stars to be estimated to within 0.1 magnitude. It is necessary, of course, for some of the stellar magnitudes to have been independently measured in order to calibrate the scale. Certain small, but complex, corrections are normally applied which make allowances for the distance of the stellar image from the centre of the plate and the colour of the star since the various emulsions have different sensitivities to red and blue light. Although this is undoubtedly a rapid method it does suffer from certain disadvantages. The variation of diameter with magnitude is very small and the homogeneity of the emulsion must be of an extremely high standard to reduce the effect of irregularities in the distribution of the grains of silver produced on exposure. When used in conjunction with a blink microscope this method greatly facilitates the discovery of new variables.

Photometric Measurements of Variable Stars

A large number of photometers have been invented for variable star work, varying in accuracy from the simple wedge photometer in which a uniform wedge of semi-transparent glass is moved over the image of the variable until it appears visually of equal brightness to a comparison star, to the far more refined instrument devised by Danjonin in which the image of a comparison star is projected into the field of the variable by means of two totally reflecting prisms, a diaphragm with a square aperture being interposed between the prisms. The area of the aperture is variable, altering the amount of light from the comparison star which enters the field. When the two images appear of equal brightness, the magnitude difference can be found from the length of the diagonal of the aperture.

Photoelectric Photometry

In the measurement of small or transient changes in brightness of variable stars, the photoelectric cell is by far the most accurate. The photoelectric effect is merely the emission of electrons from a thin film of one of the alkali metals – sodium, potassium, rubidium or caesium under the influence of light of a suitable wavelength. The cell itself is merely an evacuated bulb, the inner surface of which has had a thin film of such a metal deposited upon it at one end, forming the cathode. The anode, which collects the emitted electrons, is a wire or metal grid and is situated in front of the cathode. To facilitate the movement of electrons across the gap a potential difference of about 100 volts is applied to the electrodes and the electric charge collected on the anode is measured extremely accurately by means of a sensitive galvanometer used in conjunction with an amplifier since the resulting charge is exceedingly small, or more recently by an electron multiplier.

One great advantage of the photocell over the photographic method is that the reaction time is almost instantaneous whereas the time taken for a photographic exposure is measured in minutes or more commonly in hours. Consequently the photocell can be used to detect and follow very rapid as well as minute variations in brightness. The extreme accuracy of these measurements has enabled the limb darkening effect and the very shallow secondary minima in eclipsing binaries to be determined.

Spectroscopic Analysis

While the light curve of a variable gives us some information concerning the class of star with which we are dealing, it remains for spectroscopic analysis to provide us with more detailed knowledge of the physical and chemical characteristics of stellar atmospheres and the underlying changes responsible for the light variations. The three instruments used in the spectroscopic examination of the stars are the objective prism, the slitless and the normal, or slit, spectrograph.

The objective prism is merely a large prism having a small angle at the apex which is mounted in front of the objective glass of the camera, simultaneously recording low-dispersion spectra of all of the brighter stars in the field. No collimator is necessary since the stars may be regarded as being at infinity and the rays of light reaching us are essentially parallel. However, they are still point sources and under normal conditions would give only a narrow line on the photograph. To enable detail to be shown, the spectra are artificially widened by slow motions on the equatorial.

The slitless spectrograph is normally used in conjunction with the largest telescopes, recording low-dispersion spectrograms of very faint objects and is most suited for the study of variables when at minimum.

On the other hand, the slit spectrograph in which the light from the telescope is focused on a narrow slit at one end of a collimator, yields high-dispersion spectra of the reasonably bright stars.

Radioastronomy

So far, there are really only two applications of radioastronomy to the study of variable stars, namely the correlation of visual and radio outbursts in the flare stars typified by UV Ceti and AD Leonis and the investigation of supernova remnants such as the Crab Nebula and the Cygnus Loop. Some of the quasars too, which are extremely powerful emitters of radio waves, have been shown to be variable in light but as yet they are not classed as variable stars – they are indeed very peculiar objects about which we know extremely little at the present time.

Radio telescopes are either of the paraboloid or 'dish' type such as the 250-foot diameter instrument at Jodrell Bank or the much larger fixed dish at Arecibo in Puerto Rico which is a gigantic spherical surface of steel mesh 1,000 feet across – or they may be of the interferometer kind consisting of a huge array of identical antennae arranged at regular intervals. In general, the latter type give a higher resolution than the dish radio-telescopes.

The Future of Variable Star Investigation

As will have become evident from this survey of our present knowledge of variable stars, there are many problems still awaiting solution, problems which will require continual observation on the part of both amateur and professional astronomers. Only in the case of the eclipsing variables do we know the true cause of the light variation and even here these binary systems have proved to be far more complex than was originally thought. The dynamics of the gas streams which exist between the two stars are still far from being satisfactorily explained and the evolution of these stars presents difficulties which still remain to be solved.

The intrinsic variables provide us with a wide variety of problems still awaiting solution. It may be that in the future the many types of variable stars will be co-ordinated into an integrated sequence as has been done by Darwin in the Theory of Evolution for the animal species but this day is still a long way off and we must learn far more about these important stars before such an evolutionary scheme can even tentatively be put forward.

More immediate investigations will doubtless be carried out on such stars as the U Geminorum and Z Camelopardalis variables, particularly any evolutionary link which may exist between them and the W Ursae Majoris stars or the recurrent and normal novae; on the T Tauri stars which are believed to represent the first stages in the development of a star from the interstellar gas clouds; the flare stars for which long and comprehensive visual and radio observation is urgently required; and on the novae and supernovae which are already providing us with the answers to a great many questions concerning the original formation of the heavier elements within the galaxies.

Naturally most of this work will be carried out by the professional astronomers having the largest instruments at their disposal but much still remains to be done by the amateur. One of the questions very often asked is: Does the continued observation of such stars as the long period variables, the semiregulars and the nova-like U Geminorum stars provide useful information particularly since they have been observed for so long and many of the general characteristics of their light curves are well known at the present time? The answer always has been, and will continue to be, an emphatic Yes! It is possible that as larger telescopes come into use by amateurs, the emphasis will shift somewhat away from the long period variables to the more erratic T Tauri variables, U Geminorum stars and flare stars but there will always be plenty for the amateur to do and provided the limitations of visual work are recognised, these observations will continue to be of immense value in helping to provide answers to the many problems posed by this particular branch of stellar astronomy.

Finally, there remains the question of the discovery of variable stars. Many of the Milky Way fields have been extensively covered by photography resulting in the identification of many hundreds of faint variables but there still remain large areas of the sky which have been only scantily surveyed. A glance through the General Catalogue of Variable Stars will reveal to the reader that there are many of the constellations in which only a relatively small number of these stars have been discovered. This is not merely that these constellations are small in area. The number of variable stars known in Andromeda, Aquarius, Boötes, Hydra, Pegasus, Serpens and Ursa Major for example is noticeably far fewer than in Corona Austrina, Cygnus, Ophiuchus, Sagittarius and Scorpio although the areas covered by all of these constellations are very similar. There is therefore still much scope for the amateur in his study of variable stars.

Index